PROJECT MANAGEMENT

IN GEOMATICS

(with practical examples)

Dr. Furaha Lugoe

Foreword

This book was initially drafted as a series of papers addressing and appraising the practice of the land surveying profession. Later in 1998 the author was approached to offer a course in Project Management for students pursuing a program of study towards a degree in Land Surveying of the University College for Lands and Architectural Studies (UCLAS), now known as Ardhi University (ARU) of Dar es Salaam, Tanzania. It is therefore a result of discussion papers and a response to a special need for a text book for a course. The Course had been designed and included in the curriculum but, no text book was available and even reference material to guide the students were scanty at the time. Further, the biggest constraint was that there was no academic at the university who had majored in Project Management as such course was not offered in institutions of higher learning anywhere. One wonders why it was deemed a necessary requirement for that program of study instead of, say, hands-on internship in industry, but that was the situation.

Within three years of return to Tanzania from abroad, the author was , fortunate enough to be engaged in a diversity of projects in which knowledge and skills in geomatics/surveying were essential and project management was critical.

The first was the project to upgrade tender documents, undertake a bathymetric chart of the berth area in the Dar es Salaam Port and eventually supervise the contractor in the modernization of the port by straightening, widening and deepening the entrance channel through dredging works. In such projects progress of work is judged using bathymetric charts, obtained through hydrographical surveying, which constitute "eyes and tools" of the consulting engineers.

The second project was the Divestiture of sisal estates of the Tanzania Sisal Authority (TSA). Over the years the estates had been neglected since the market for sisal had been depressed and the State-owned TSA could not balance its books. In a paradigm shift away from state control of the sisal industry, the government decided to sell off the estates to willing buyers. When this decision was made, the reality came out with regard to tenure security of the estates' lands that most of the estates were also encumbered in title. Villages located close to many of the estates had invaded and used the estates for crop agriculture. Due to many years of fallowing and neglect local government had unofficially sanctioned this alternative land use change to adjoining villages and villagers had invaded the estates and used them as private lands. The Central Government then decided to remove the areas occupied by villagers from original titles in favour of the villages and villagers and sell-off the remainders. In this case, farm survey were needed in order to excise the lands as directed by regional governments, resurvey the remaining lands and obtain new titles in readiness for sale. Thirdly, in 2001/02 the author was involved in the Public Expenditure Reviews (PER) studies for the lands sector. These studies had become a pre-budget requirement, especially for priority sectors of Tanzania's economy or sectors with priority outcomes prior to budgeting. Such involvement continued for two more years during which a thorough study of the services of the various divisions in the Ministry of Lands, Housing and Human Settlements Development was undertaken and unveiled many problems, constraints, opportunities, etc in the sector. Professional matters especially of land surveying and mapping branch and their bearing on other departments and clearer performance of the Ministry were brought to light. It was at this point that request by the Land Surveying Department at UCLAS was made to the author to teach the Project Management course part-time. The experience gained so far provided material for the course and the first draft came out at about this time.

It must however be mentioned that upon returning to Tanzania the land surveying profession was not enviable. Private practice was itself depressed for a number of reasons: Firstly, there was literally no market for the services of private land surveyors as Government surveyors did all the jobs. But as discussed elsewhere in author's publications, their outputs were also very low and far from satisfying the demand. Secondly, the national economy was not performing well and budgeting was not in favour of land development services. Purchasing survey equipment was seen as an extravagant expenditure.

Thirdly, in the private sector the foreign exchange was a big problem before the government decided to label survey equipment under the same group as computers which were then favourable. Fourthly, few graduate land surveyors were registered and even fewer were licensed to practice for gain. Licensing was a Government prerogative but in practice government surveyors were competing for the same clients as private surveyors. Yet the surveying fraternity in the private sector had no voice as the institution of surveyors of Tanzania (IST) was as good as non-existent.

In the first five years most of the author's time was spent on professional matters as a way of making the ground to practicing for gain. The author served several years as President of the IST. In essence the following material were prepared and discussed:

1. Lugoe, F. N. (1995). "On Land Surveying Professional Practice in Tanzania" Occasional Paper, Topo-Carto Consultants Ltd.

2. Lugoe, F. N. (1996)"Contracts in Professional Practice" – A lecture in professional practice. Occasional Paper, Topo-Carto Consultants Ltd.

3. Lugoe, F. N. (1999). "The Business of Surveying the Land" – A lecture in professional practice. Occasional Paper, Topo-Carto Consultants Ltd.

4. Lugoe, F. N. (2000). "Why Land Surveyors Cannot Make a Living from their Profession in

Tanzania – The Poverty Syndrome."IST. Occasional Paper, Topo-Carto Consultants Ltd.

5. Lugoe, F. N. (2001). A Proposed Agenda for the Institution of Surveyors of Tanzania (IST) for the year 2025. Proceedings of Symposium on the IST Vision, 2025. IST Annual General Meeting and Symposium, 2001

6. Lugoe, F. N. (2002). Economics of Land Surveying. Researched Notes for Students in the BSc Geomatics Programme. T-CC Map and Print Centre. 2002

It was in author's view that the land surveying profession needed to focus on professional practice issues underlying the poverty syndrome that had manifested itself among graduate land surveyors. But, also a good knowledge of contracts in the business of land surveying was necessary. Understanding the economics of land surveying, as may be different from other professions, was also of utmost importance as many could not differentiate between working for gain and providing a service, even the pricing of works were leading many into loss-making. The call to offer a course in Project management for three years concurrently gave the author the rare opportunity to set the foundation of contents of such a course based on examples in land surveying albeit, personal examples which could be expanded upon in the future. The author recalls writing a set of three documents to serve as lecture notes to students under the titles:

1. On Land Surveying Professional Practice in Tanzania (1995)
2. The business of surveying the land- a lecture in professional practice (1999) and
3. Contracts in Professional Practice- a lecture in professional practice (1996)

These culminated in a compounded document titled "Economics of Land Surveying.- Researched Notes for Students in the BSc (Geomatics) Program" published in 2002. This document has been included as a chapter in a recent book entitled; "Spatial Data Acquisition, Display and Use" so as to convince students to get used to project management as an integral part of learning in these sciences. The author has however, found it of utmost importance that theory be complemented with practice and have hence devoted a chapter on four projects covering activities of his company, Topo-Carto Consultants, where the executioners or contractors and in which the contribution of land surveying knowledge and skills are essential to their execution and hence successful completion.

I recommend this book to students in the subjects of land surveying and geo-information but, also urge the old timers to read it as they continue to practice their professions and provide advisory services in projects involving their knowledge and skills to a wider audience.

NOTE

In this book the words geomatics and land surveying will be used interchangeably. In order for this topic to be well understood, it is important that the terminology in it be defined first. Project management is the sum total of technical and social economic activities made in ensuring positive outcomes when resources are invested. Both basic micro and macroeconomic aspects of economics shall be presented first to put geomatics in the perspective of financial investment and in the context of landed projects.

PART 1: THEORETICAL CONCEPTS

Chapter.1: GEOMATICS OR LAND SURVEYING PROJECTS

By project we understand any scheme for investing resources. Such a scheme will be such that it can reasonably well be analyzed and evaluated as an independent unit. This definition is quite broad guided by whether or not the schemes independency of other undertakings can be established with respect to inputs and outputs. It is generally possible to establish units within a project with enough levels of independence as to constitute smaller projects within the main project.

Land Surveying or geomatics, in modern terms, is the science of collecting spatial data and transforming that data into information about the land and using that information to enable or facilitate outcomes in other fields of knowledge. Raw data is collected by making measurements between objects and land features or by electronic capture on the relative positions of such objects. The raw data are then transformed into data that is descriptive of the land and its features to produce land data. The land data is further processed by computations and transformed into land information displayable in maps, charts, graphs, etc. Geomatics or Land Surveying is therefore a measurement, computing and information science also known as geoinformatics. It is comprised of sub disciplines, and suffices to mention geodetic, topographic, photogrammetric, cadastral, engineering and, in its broad sense, hydrographical surveying as the subjects within it. It is so closely related to geodesy and cartography that it is assumed that the two are its essential parts.

Having described what a project is and what constitutes geomatics or land surveying, we are ready to define what constitutes a geomatics/survey project. In the simplest terms a Land survey project is an investment scheme whose object is to acquire and present land data and information or is a scheme aimed at employing such data and information is another project, such as engineering or defence project. The products of a survey project are therefore data and information about the landscape and its features such as:-

- Positions and position lines,
- Heights and height differences of points or between them,
- Scaled survey plans and bathymetric charts of the topography,
- Geo-referenced cadastral records,
- Thematic and topographic models,
- Profiles, cross-sections, earthwork quantities, lake capacity measurements,

- Information on structural stability, catchments, drainage, river flow, etc.

A land survey project, like many other projects, can be broken down into smaller projects as stated earlier. The smaller ones are designed to work without each other but are small enough as to enhance workmanship and cost effectiveness in their management. Smaller projects are in many ways preferred to larger projects especially where it is not easy to find a variety of skills under one implementing agency of company. A contractor who does not have all the capabilities is normally allowed, under contract, to engage a subcontractor who would only be answerable to the contractor and not the client. This situation sometimes causes conflicts between contractor and employer/client or conflicts in project scheduling. Alternatively a project may be decomposed into smaller ones, that can be handled with a single contractor specialized in the smaller component.

Example of Decomposing a Project: Taking an example the Topographical Mapping of the Mwanza – Geita Block undertaken from 1991 to 1996, one could easily identify the following seven units:

- Aerial Photography,
- Network Densification,
- Photo-control surveys,
- Aero-triangulation,

- Photogrammetric machine plotting,
- Cartographic Map preparation,
- Map production and printing.

But, as a project it could easily be decomposed into four independent and smaller projects each with a contractor and manager. These could be: (i) Aerial photography, (ii) Densification and photo-control surveys, (iii) Aero-triangulation and machine plotting, (iv) Map preparation and production.

Micro/Macro – Economics of Geomatics

A dictionary of economics and commerce (Hanson, 1977) describes economics as a social science concerned with three issues:-

- The many problems of production but not the technique of production. These include the scale of production, whether firms carry out commodity production from start to finish or undertake only single processes,

- The distribution of the national income, which involves the study of money and the determination of prices,

- A description of the working and functions of economic institutions such as central and commercial banks and the stock exchange.

The first issue is a concern of the surveying industry and to some extent part of the second issue. Economics can therefore be summed up as a science that studies human behaviour as a relationship between ends and scarce means with alternative uses.

It is now possible to understand our heading from these few clarified terminologies. Economics of Surveying looks at surveying and surveyors as a production and distribution institution that includes: (i) National survey authorities such as the surveys and mapping division, SMD, in Tanzania and similar agencies elsewhere (ii) Parastatal organizations, (iii) Companies limited by shares, (iv) Partnerships, (v) Non-governmental organizations (NGO), (vi) Solo practices. And, their day to day productive activities, the products of which have already been listed under 'geomatics projects'.

The subject matter of economics is conveniently studied from two subjects or branches, namely microeconomics and macroeconomics.

Microeconomics is defined in economics textbooks as a branch of economics that is concerned with individual firms, their outputs and costs, the production and pricing of single commodities, wages of individuals, etc as distinct from their aggregates. Similarly, Macroeconomics is concerned with aggregates of production, consumption and income of a community as a whole. It considers the relationships between volume of employment, total amount of employment, total amount of saving and investment, the national income, etc. It is apparent therefore, that at the central of both micro and macroeconomics lie processes of production of commodities or products. In this context surveyors/geomaticians are as much producers of consumable services and products as say, agriculturalists.

Production and Products

Products can either be primary, secondary or tertiary. Primary products are raw materials that do not require further processing but may be taken as a basis for the production of secondary products. In land surveying the primary products are rarely useful to a client save as a basis of further production. These are: (i) The outcome of field works in the form of abstracts of measurements, whether digital or manual, contained in completed angle sheets (A-S), measurement sheets (M-S) and detail sheet (D-S) often submitted after field operations, (ii) Description cards of erected monuments and, (iii) Field sketches drawn as a guide to plotting.

Secondary products are manufactured and hence, processed products. In land surveying these are obtained from measurements and data capture only and are the results of adjustments, computations, and data or information displays. To this category the following can be identified: (i) Positions of points in the form of coordinates, (ii) Position lines in the form of bearings and distances, (iii) Heights and height differences, (iv) Survey plans, bathymetric charts, thematic and topographic maps, whether presented in hard copy format or digitally, (v) Profiles, cross-sections, areas, volumes etc, and (vi) Displacements, height differences, etc.

Tertiary products are essentially personal services rendered. These require that the surveyor be equipped with information and expertise to use that information productively. The services include: (i) All forms of evidence provided on the basis of survey data in the solution of say, boundary disputes, (ii) Studies requiring knowledge of surveying, (iii) Expert information given in the adjudication of title process, (iv) Site services in setting-out works, (v) Alignments of structures and buildings, (vi) Route location surveys, (vii) Consultancy and advisory services in policy and strategic studies, etc.

Survey practice for gain requires that each product in the production process be identified, categorized, marketed, costed and priced. This is particularly important in a contract whereby the client requests the services of a land survey firm without presenting elaborate terms of reference (ToR). It is the duty of the bidding firm to explain, during contract negotiations, the expectations of the client in terms of survey products and services as reflected by the ToR. The type and quantity of the product determines the magnitude of the production process and particularly, identify the resources to be invested. This is a very important factor in project design and management.

Chapter 2: PROJECT MANAGEMENT

The quality of products in the production process and the balance between resources and products is determined by, among others, the quality of management of the project. Project management aims at project implementation in accordance with the design, using prudently, the resources identified. Imaginative and flexible project management can correct even faults in the design and turn the project into a success. This section will discuss aspects of project management principles used in surveying/geomatics and which are a key to successful production.

The Over-arching Principle (OP)

This is the principle of working from the whole to the part and from the large to the small and not the other way. All survey operations are guided by this principle. The OSP guides design and execution of surveys both large and small. In a cadastral survey of urban plots for example, control is established to cover the area of the survey. Then the block is surveyed and finally the block is divided into individual plots.

Principle on Procedure (PP)

As a rule, primary products in surveying are obtained in the field and secondary products are processed in the office. The tools of labour are categorically different and unique in each case. The key principle on survey procedure is therefore the principle of categorizing all survey work as part of field or office operations. To the group of field works can be identified the following: Reconnaissance; Testing and adjustment of instruments and equipment; Monumentation; Bush clearing; Field sketching; Measurement of distances, directions, levels, GPS ranges, GPS range differences, gravity, meteorological parameters; Assessment of observations; and Compilation of measurement abstracts.

The following operations belong to the group of office procedure: Project preparation including design and its pre-assessment; Choice of measurement techniques; Selection of instruments and corresponding tests/checks; A priori data search; Final checks on submitted data for errors and with respect to the design; adjustment computations; Storage and data base management; Plotting and draughting; Graphical representation and other forms of data/information display.

A good knowledge of this key principle on procedure is essential to everyone engaged in land surveying particularly, those dealing with project planning, scheduling, execution and particularly project costing.

Production Principles Based on Expected Data Accuracy (EDA)

Most of the work undertaken by land surveyors is aimed at determining position of features in a predetermined survey coordinate system. Many of the remaining secondary products are derived from these positions. Positions can be either absolute or relative but even the relative positions require that somewhere there be established absolute points. Why is it that although expertise exists for determining absolute positions, it is seldom that absolute positions are determined in practice and all surveys deal with relative positions? The following discussion explains the situation and derives four production principles that are EDA based.

Absolute positions can be established astronomically or using satellite based systems particularly the Global Positioning System (GPS). The astronomic and geodetic coordinates derived can be transformed into place coordinates using conformal projections such as the Universal Transverse Mercator (UTM) system. The two have serious shortfalls in absolute positioning.

Errors in the coordinates for absolute astronomically determined positions are large. An error in latitude of

0.5 seconds of arc corresponds to 15m in UTM coordinates. Achieving that kind of precision for a single point will require: (i) Long periods of waiting for a clear sky, (ii) Several days of night time stellar observations, (iii) The highest class of instruments (hence very expensive) for angular measurements and timing, and (iv) Good logistical support, etc. It is clear from the aforesaid that products based on this method of production will be too expensive to find a market.

Equal well, high accuracy in GPS's absolute positions does not come easy although with effort and resources 1-2m accuracy in coordinates is now possible. It is however achieved using: (i) Top of the line GPS receivers with dual frequency, (ii) The P-code, (iii) Precise ephemeris in computations, and (iv) Proper modeling for errors in the ephemeris, clocks, ionosphere, troposphere, multipath, measurement noise, computations, etc. Under normal circumstances, the C/A-code and broadcast ephemeris are used instead, bringing the capability of the GPS absolute coordinates to nearly the same level as the astronomically determined ones. Consequently the following four principles are, as a rule applicable in surveying:-

The Principle of Relative Positioning

Due to a number of reasons stated below, surveyors do not, as a matter of principle, survey for absolute coordinates in the implementation of survey projects, but most survey projects are achievable through relative surveys. Absolute positions are surveyed only in the definition of the datum.

The Principle of Networking

To ensure uniformity of accuracy, a geodetic control network (GCN) system is, as a matter of principle, used to transfer absolute positions to other points of the geodetic datum. The errors embedded in the absolute positions are systematic in nature and do not substantially affect the relative positions of the other points in the GCN. Further, the GCN system is, also as a matter of principle, categorized into several orders according to standards of accuracy that decreases in the lower orders. The orders in the context of Tanzania are:- Primary, Secondary and Tertiary networks. The tertiary order can be followed by standard traverses particularly in urban centers and remote areas with little development activity.

The Principle of Densification

The lower order surveys constitute a densification of the Primary order of the GCN. Surveys are as a matter of principle to be tied to high order surveys as part of the densification process. The categorization of surveys into a hierarchy of several orders is very much also a response to the OSP defined above. The Primary points normally cover the territory though sparsely.   Lower orders may fill up the gaps and further densification will bring control down to the user.

Chapter 3. PRODUCTION AT INSTITUTIONAL LEVEL

The institutions involved in the production process of survey products have been identified earlier. The organogram for the institutions considers the liner relationship between management, administrations, and production groups. Bigger organizations are more streamlined in functions of individual personnel. Smaller ones assign more functions to a single employee instead. The table below shows a simplified organogram of survey production institutions, comprised or only two departments namely the technical and administrative departments.

It was stated earlier that producers of survey products can be as large as SMD funded directly by the state, or as small as solo practitioners funded by individual persons' resources. An attempt is made below to list task that are normally performed y various institutions in the production process.

The Surveys and Mapping Division deals with surveys of a large extent such as: (i) Definition of the horizontal and vertical datum, (ii) Establishing the Geodetic Control Network, (iii) Establishing the Geodetic Levelling Network, (iv) Densification of the geodetic networks to a level that is economically convenient for various applications such as township mapping, (v) Cadastral surveying and record keeping,

(vi) Topographical mapping at various scales, (vii) Hydrographical surveys and charting of the exclusive economic zone (EEZ) and lakes, and (viii) Maintenance of territorial boundaries. SMD operations are guided by legislation particularly, the Land Survey Ordinance (Cap 390), the Professional Surveyors Registration Act No.2 of 1977, and government policy. Some of the work can be contracted out to private firms.

From the economic point of view the responsibilities of SMD yield mostly tertiary products. This is apparent from the above listing where datum definition, establishing and densfying networks, boundary and title adjudication, as well as maintenance of international boundaries are accomplished in the provision of an infrastructure that facilitates other surveys, mapping, defence programs, engineering constructions, deformation monitoring, land parcel registration and titling, geo-referencing satellite imagery, planning of settlements, etc, and as an expert witness on the national cadastre. These products cannot easily be provided by private undertakings primarily as they constitute a public service. The price responsibility of SMD can therefore be summarized as that of developing and maintaining a survey and mapping infrastructure for the national economy.

The surveys and mapping division produces secondary products in the form of maps and charts, aero-photo products, cadastral plans and records as well as positions for referencing aero-photos in the mapping process. There products can also easily and more economically provided by the private sector where competition exists and therefore a greater chance that costs could be brought down substantially.

Parastatal Organizations deal with the following two. Firstly, survey of lands under their jurisdictions as directed by management, and maintain survey records of work done for their respective organizations. Production, in the Private Sector, is undertaken by limited companies, partnerships and solo practices. These concerns deal with:

•	Various surveys for individual clients and firms,

•	Cadastral surveys of private lands or as contracted by government,

•	Other surveys as contracted out through SMD or engineering project managers,

•	Consultancy and services in land related issues.

Non-governmental organizations produce any of the above as directed by their objectives and interest of donors. The Private sector and quasi-governmental surveyors produce mostly secondary products such as already enumerated but produce a substantial amount of tertiary products in their consultancy services and advice particularly in servicing the land tenure system and in engineering project.

Chapter 4: COSTING AND VIABILITY

Costs in General

By cost of a product one understands the sum of all the payments to the factors of production engaged on the production of that particular product. The term cost has meaning only when related to the output and is often referred to as cost of production. Costs can be categorized as prime costs, supplementary costs, fixed costs, variable costs shut down costs and marginal costs

Fixed costs are those costs, which do not vary with the size of its output. It may include: Money set aside for depreciation of field capital i.e. buildings, plants, field equipment, and Salaries of managerial and office staff if the number is maintained whether or not production changes. These are variable costs together with the cost of administration, which only in the short run is to be regarded as a field cost. However there are few costs of a non-variable nature in production. In the long run there are no fixed costs and all costs are variable.

Prime costs include all the variable costs together with the cost of administration, which only in the short run is to be regarded as a fixed cost. A firm will be required to shut down when its income fails to cover its prime costs. A shut down cost is therefore attained when revenue equal to the prime costs.

Supplementary costs are the remainder of the fixed costs when administrative costs are removed.

Marginal costs are the addition to its total variable costs that must be made to increase its output by one unit per period of time. Any convenient unit of production may be selected, by a firm, for use in the calculations. Fixed costs do not affect marginal costs. Often, when marginal costs increase, it is accompanied by an increase in output.

Costs in Geomatics Operations

It has been stated that projects involve the investment of resources. Let us now examined the kind of resources that are common to production in a survey project. To begin with let it be stated that the resources involved in survey production include: (i) Skilled and unskilled personnel, both managerial and workers, (ii) Equipment for acquiring raw data in the field or data capture, (iii) Equipment for processing, storing and displaying secondary products, (iv) Equipment for logistics such as vehicles and field gear, (v) Buildings and furniture for the office, (vi) Services rendered to facilitate quality production such as calibration lines, equipment-hiring outlets, the internet, external data processing centers, etc. In land surveying production there are several types of costs can be identified. These are:-

i.   Salaries and wages, these are the pre-tax gross regular cash payments made to each employee without considering bonuses or premiums,

ii.   Social charges are the non-monetary benefits of all staff and workers, which include pension, medical expenses, life insurance, annual leave, housing, car allowance etc,

iii.   Office overheads are business costs in the office that include office rent, water electricity, equipment maintenance, temporary support staff payments business promotions, etc,

iv.   Allowances are bonuses and other payments made as an upkeep to personnel when working away from duty station and include per diem, board and lodging, field allowance for working under circumstances of particular hardship or outdoors,

v.   Project overheads are the means of production and include; plant and equipment whether hired or owned, transport, accommodation and furniture for field office, staff accommodation unless board and lodging has already been considered, room for storage of equipment and material, field supplies secretarial facility for field office, mobilizations, and taxes.

Both fixed and variable costs can be identified in these five groups.

Supply and Demand

The importance of cost in economics lies in its influence on supply, where a rise in cost tends to curtail supply and vice versa. Cost of production also has a direct bearing on the price of products. Cost of production is made up of the prices of the various factors of production. The cost of each factor, as is the cost of the product itself, depends on what it can earn elsewhere, i.e., its demand. The price of a survey product is therefore simply, the amount of money that has to be paid for a survey product or service. Once determined a change in the price of a commodity does not come about by itself, but because sellers are prepared to accept less and buyers are prepared to pay more.

Under imperfect conditions of competition a firm may fix its prices of the basis of its costs. The profit margin will depend upon the state of demand. Supply and demand go hand-in-hand and obey the following four laws:-

i.   Price tends to equate the amount that sellers are prepared to offer for sale and the amount that buyers wish to buy (willing seller - willing buyer),

ii.   Usually a larger quantity of a commodity will be demanded at a lower price than at a higher price; and a larger quantity will be offered for sale at a higher price than at a lower price,

iii.   An increase in demand tends to raise the price and to expand the supply; a decrease in demand tends to lower the prices and to contract the supply,

iv.   An increase in supply tends to lower the prices and to expand the demand; a decrease in supply tends to raise the price and to contract the demand.

Product Value

The value of a product is determined by the cost of production for, under perfect competition the price of a product will tend to equal its marginal cost of production which, in turn, depends partly on demand. The TI4100 GPS receiver manufactured in 1984 was the first multi-channel, dual frequency receiver with geodetic precision. These were manufactured as a response to firm orders from clients. One set was sold at a price as high as US$ 200,000.00 (about ten times the value of the state –of –art GPS receivers of today). In determining its value, consideration was made for a number of factors, namely: (i) The investment made by Texas Instrument inc. in research and development for about five years, (ii) The demand for such GPS receiver on the marked and, (iii) The low level of supply by a single manufacturer in the world.

Fee-for Service

A fee is simply understood to be the price of service payable usually to professional person as a return for every item of service rendered. It may be provided in addition to or in lieu of a salary. It has the advantage of encouraging professionals to maximize productivity i.e. the number of procedures derived from given inputs, such as a medical doctor will be attracted to perform more operations if paid a fee per operation rather than a monthly salary. Similarly, a land surveyor will strive to survey more plots per unit time if a fee is paid per plot output.

A fee also encourages the professional person to expand sale of units of service beyond those carried out under other forms of re-imbursement, and to perform unfamiliar procedure in order to gain a fee. A lecturer who teaches more hours will be encouraged to continue teaching more than the acceptable average and to attempt teaching new courses if a fee is payable to the extra work done. Most accounts personnel in Tanzania work overtime (After hours and on week-ends) because they get paid for the extra hours. If an employee's salary therefore varies with the number of procedures carried out in some previous period than that payment ceases to be a salary but a fee.

A consultant, who is normally not in the employment of a client, is not paid a salary but a fee for services rendered. Bothe the consultant and client will agree on both the qualitative and quantitative aspects of the engagement and on a fee to be paid per unit time before the service is to be discharged. The fee is set at the normal salary that would be payable elsewhere for the qualification and experience of the consultant, the social charges, office overheads and a profit. The fee payable in this regard is called the man-month-rate (MMR). If the engagement requires the inception of other project overheads such as equipment, transport, accommodation, etc., then the consultant must indicate to the client of the desirability to be included into his fees a component of these overheads as reimbursable. The consultant must indicate all these figures as elaborately as possible at a contract negotiation session with the client including the desired mode of payment and re-imbursement.

Cost/Benefit Analysis, CBA

Cost benefit analysis, also called cost-effectiveness analysis is a technique for the evaluation of an existing situation whereby the social cost of a project is considered in relation to the benefits it confers on the community. The technique originated in the USA in 1930 by engineers on water related projects and is now applicable to almost every field where the output has a market price and the social benefits can quantitatively be measured. Difficulties for the application of CBA are still experienced in education, health and defence for the same reasons.

CBA is normally used in considering say a site for a new airport, dam, highway, etc. The process of land delivery of a built up urban area for settlement is also a subject of CBA. Others may include squatter upgrading, demolition of squatters, etc. Benefits can be obtained in the form of profits, which in capitalist thought is a measure of the social and private gains that the community gains from a project. Profits are an important signal for guiding investment decisions if expenditure measures closely the social cost and receipts measure the social benefits. CBA therefore seeks to itemize the social costs and receipts related to such major projects.

In building a dam at a river gorge, a reservoir is born that covers a substantial area upstream. The area covered with water is usually a centre of livelihood for a local population who are to be compensated for loss of property and investment, and a loss of flora and fauna and be resettled,. The benefits of the resettlement and land acquisition would be: communities gaining a possible source of water for irrigation, power generation, domestic and industrial supply, fishing, tourism, employment, etc. The costing of both expenditure and receipts will be done as part of CBA prior to deciding to proceed with the project. The difference between receipts and costs measured at accounting prices is, therefore, most appropriately called social profit. The level of social profit will determine the adoption or rejection of the project.

Chapter 5. **COST AND FEE SCHEDULE (CFS) OF THE INSTITUTION OF SURVEYORS OF TANZANIA (IST), 2001.**

Background

Land Surveyors are in the business of surveying and mapping. As it is with any business venture, the services are offered for gain. This is a clear statement of fact that must be underscored. Gain is the opposite of loss and, I must add, that there is no gain in a brake even situation either. Working for gain means being able to pay for all operational cost in a project and, in addition, make a profit if one is to sustain the practice and allow for growth. In this regard, working for gain means that the client pas for the full cost of the services requested and rendered. It needs no reminder but in a market economy, and particularly in the private sector, subsidy does not exist and was not presumed in the CFS adopted by IST at the 1999 annual general meeting.

Efforts to change the Practice

In March, 1995 a group of Land Surveyors had an audience with a representative of the Director of the state Survey and Mapping Division and brought the issue of pricing survey services into debate which continues to date. In February 1998, in a speech to the Institution of Surveyors of Tanzania (IST), the

Director revisited the issue. He pointed out that in his view "among the disabilities of many Land Surveyors" are the "incapability to put a price tag on the labor involved in the business", failure to set up a uniform pricing model" and "general lack of investment in the profession". The Director went on to ask why IST fails to put in controls for the business and use the national council of professional surveyors (NCPS) against those who prostitute the profession".

These visionary ideas in the continuing debate on "returns for labor" was taken a step further by the IST's extra-ordinary AGM of 1998 by adopting a resolution in favor of a CFS (see annex).

Project Costing Models

In preparing the CFS the IST Council examined two costing models for the land surveying industry in Tanzania with the following views in mind:-

☐ By seeking professional land survey services and the offer being accepted, the two parties enter into a binding contract whether written or verbal. The obligation of the Land Surveyor is to provide the promised end product on time whilst that of the Client is to pay the money, also timely.

☐ The cost of execution of all surveys necessary to compile and present the end product is for the account of the Clint who pays professional and other fees as agreed. In other words, there is only one source of payment where the Land Surveyor will direct all his invoices.

The first principle is often taken for granted and hence acceptable although its implementation is taken lightly and possible consequences are not carefully examined.

The Rate Model

The rate model sets the cost of a unit survey activity that, in this case, is the cost of positioning a survey point. Positioning a survey point is a sum total of all survey operations deemed necessary and sufficient to accurately define the co-ordinates of an appropriately monumented and witnessed survey point including, and assessment of the positioning accuracy and ensuring that it meets quality assurance criteria for the intended purpose.

A survey point is often an end product in itself and clients understand it more than other activities when related to their needs such as plans, boundaries, profiles, cross-sections, etc.

The rate model is technology dependent, meaning that the cost will vary with adopted methodology and instrumentation that are largely responsible for the speed and accuracy of the surveys. It does not mean, however, that the cost of a project will increase with choice of a more modern technology for a given project rated costlier in the model. Modern technology, for instance, enables an easier and faster link of new to old points, past obstacles in inter-visibility, etc.

The Time Model

The primary factor in the time model is time expressed in hours or weeks or months and the monetary value of the time that a Land Surveyor works for the Client. The Land Surveyor starts with a comprehensive study of the project and decides on both the quality and quantity of resources to be deployed.

The resources for the task include personnel of all categories and the time that each will spend on the project. The remuneration of each person will then be decided, based on each one's salary, to arrive at what, in contract language, is known as man-month rate (MMR), even if hourly or weekly rates are to be forwarded to the client.

The MMR is computed as a sum of a number of factors. These are; a monthly salary (50%), social charges (13%), overheads (17%), field allowances (10%) and profit (10%). The figure in parentheses is typical of percentages anticipated and in the market place this figure is usually a give or take of up to 5% of these values. In a contract, the Land Surveyor is required to justify the percentage point to be entered into the contract. It is in the interest of the Land Surveyor to ensure that the rates are realistic and that he will make a profit from the project but also remain competitive with others who would bid for the same project. May I briefly elaborate on the factors comprising the MMR.

i.   ***Salary*** – is the pre-tax gross regular cash payment made to each employee in the survey party at the time the project is conceived without any bonus or premiums that may be paid on assignments away from duty station.

ii.   ***Social Cost*** – is the cost to the Land Surveying company of all non-monetary benefits for all staff in the survey party. It includes such items as pension, medical expenses, life insurance, annual leave, leave at the end of an assignment, staff replacement of staff on sick leave or vacation, housing, bus allowance, etc.

iii.   ***Social Cost*** – is the cost to the Land Surveying company of all non-monetary benefits for all staff in the survey party. It includes such items as pension, medical expenses, life insurance, annual leave, leave at the end of an assignment, staff replacement of staff on sick leave or vacation, housing, bus allowance, etc.

iv.   ***Overheads*** – in this case are the business costs of the Land Surveying firm that are not directly related to the execution of the project being designed and include office rent, water, electricity, equipment maintenance, cost of supporting staff such as temporary employees, secretary, messenger, etc, and business promotion costs.

v.   *Field allowance* – is payment made to field party members when working in circumstances of particular hardship outdoors and often requiring the surveyor to work off and longer hours.

vi.   **Profit** – has been described as the cost of risk taking and is the basis of sustainability and growth in the land survey or any other business.

The time model would be incomplete if only the MMR was to be negotiated with the Client, without due consideration to the cost of the means of labour in accomplishing a given task.   Another major item therefore is the logistics often referred to as the project overheads as different from office overheads discussed above, or re-imbursables.   Perhaps the last terminology is more descriptive in the sense that these are expenses borne by the Land Surveyor and reimbursed by the Client.   In a contract, it is imperative that these be known and agreed upon by both parties before entering into a contract.

Project Overheads

Overhead expenses in land surveying consist of the following expenses: Plant and Equipment, Transport, Accommodation, field Office, Mobilization, draughtsman ship and other. An attempt will be made to include all known project overheads into these seven groups.

☐ ***Plant and Equipment*** – the Land Surveyor lists all equipment to be used on the project whether own or hired and the cost to the Client for their deployment on this project. The list must be comprehensive in the sense that all equipment will be listed. The cost of own equipment will be what the Land Surveyor would hire them out on the market for the duration that the item is deployed on the project. The cost of hired equipment would be the market rate plus 15%, also for the number of days that the equipment is not lying idle. One can see that hired equipment will be more expensive to the client that owned and a Land Surveyor with own equipment would be more competitive than otherwise. Further, a wise choice of equipment can speed up field operations and reduce office work considerably thus enabling the Land Surveyor to reduce project contract time and consequently, the project cost or contract sum. The Land Surveyor is not supposed to make money on equipment and the Client reserves the right to pay only those expenses, which can be supported and verified. It is unethical to list an item and not deploy it, yet charge the client for it.

☐ ***Transport expenses*** is the cost of deploying vehicles on the project. Again, only vehicles that are necessary and sufficient for the job will be included, whether own or hired, at market rates and a 15% mark-up for the latter. The rates for owned vehicles must take into consideration cost of fuel and maintenance and payments to drivers. Rates for hiring vehicles on the market are very high and to be competitive the land surveyor would better charge mileage as an alternative.

☐ ***Accommodation*** in the case of projects is three fold embracing office accommodation, staff accommodation and storage. Often the project location is away from home office and the need to open an office and accommodate personnel is compelling. Sometimes there is a need o set up a field office in the same town. Even when the Land Surveyor still operates from the same office as before, there may be a need to expand due to increased numbers of staff and work. Land Surveyors report to the use of camping gear and cook own meals so as to cut costs and the competitive. If security guards are hired, their payment is also included in accommodation expenses. Staff accommodation includes full board lodging for staff and out of pocket expenses. Storage refers to a secure equipment store for the project.

☐ **Field Office Expenses** are additional office accommodation and operating expenses due to the inception of a new project in the operations of the surveyor. These include expenses made towards rent, water and electricity, furniture, stationery, secretarial services, phone and fax, tea and packed lunches, computer-time and printing.

☐ ***Mobilization*** Includes the cost of reconnaissance, moving personnel and equipment to the project site and their return on completion of the task, payment for insurance of various kinds, payment of statutory fees where applicable, procurement of beacons or other monumentation. It is quite possible that more vehicles would be required here than at any other time during project execution.

☐ ***Draughtsmanship*** includes the costs of preparing hard copy displays for the client. This is usually a lump sum that includes payments for labor and material.

☐ ***Other Expenses*** not covered here are possible statutory fees, report writing (for a required number of copies) and presentation. Land Surveyors must pay attention to taxes demanded of them, in particular pay as you earn (PAYE), value added tax (VAT), retention tax, community tax, etc and the costs of licenses.

☐ ***Profit and Contingency***

We have so far dealt with two elements of costing based on the time model namely labor cost and the project overhead expenses. We shall conclude by introducing two more elements that by now may be obvious to the reader, namely, profit on the project and contingency.

Company Profit is calculated after all other elements have been considered and added up. It will lie between 12% and 20% of the sum. As all other cases the Land Surveyor must be able to convince the client that the selected percentage is appropriate.

Proposed Project Cost

Number	Cost Item	Proposed Sum
1	Project Overhead Expenses	
2	Cost of Labour	
3	Company Profit	
4	Discount (if any)	

Contingency money is a part of every project. This money is not there for the taking but will only be used with the permission of the Client on unforeseen aspects of the task, which were omitted from the contract, say, by error. The Land Surveyor must insist that the Client set aside a sum of money amounting to 10% - 20% of the contract sum as contingency money that will be readily available for additional minor assignments on the project known in contract language as variations. Contingency is not included in the computation of the project cost, but must be stated in the contract.

Factors Contributing to High Project Costs in Land Surveying

The cost models used by IST are comprehensive in the sense of incorporating all the main factors that may, if not considered, lead to loss making. The same may lead to, either overpricing or under-pricing if not optimized. These situations often occur when ground-work is inadequately done. Assuming that the project is well 'coasted', there are still a number of factors that will always lead to high costing in the Tanzanian context. I would like to discuss the major contributory factors here, namely, the paucity of control points and adverse field conditions, the composition of the survey team and the cost of survey technology on the market.

In Tanzania, the paucity of existing control points and, to a large extent, outdated based maps and aerial photographs will unfortunately keep costs of survey projects very high for the foreseeable future unless the State starts to meaningfully invest in network densification and map revision. The paucity of survey control is also closely related to speed of surveys and invariable to the cost of survey projects.

Other field conditions adversely affecting speed of survey operations are inter-visibility of points to be surveyed and their separation accessibility, the weather and seasons. These factors compel us to use pass points which are neither for the Client's nor Land Surveyor's benefit but the Client has to pay for their cost anyway. They also require bust clearing and temporary monumentation. Field teams spend expensive time to reach remote points and those on difficult terrain or away from roads. Surveying is a hard outdoor operation that may be interrupted or suspended during rain and storm in search of shelter. In torrential rains the roads may be destroyed and project sites inaccessible.

i.   The Paucity of Control Points

The Government has on several occasions been appraised of the problem of paucity of control points in both urban and rural areas. It is a critical shortfall in the activities of the activities of the Surveys and Mapping Division that practically stopped three decades ago. A1995 needs assessment study commissioned by the Government also recommended particular attention to this issue. One wonders, why this state of affairs is unique to Tanzania, a country that has, of late, made a substantial technological leap with a GPS arsenal that is probably second to none on the continent. May be to elaborate on the issue it is worthwhile to digress and invoke and analogy in a way of a question for us all to ponder. What would be the cost of manufactured products if factories were to produce own electricity, pump and purify own water, use own telecommunication units! If that sounds primitive and wasteful then, what about the current practice of each surveyor setting up survey points and thereafter discarding the same without any documentation and preservation.

Land Surveyors are now acting like factories that establish own infrastructure each time a new survey is commissioned and disband it without trace thereafter. Often the infrastructure is more expensive than the factory itself. Imagine the cost of our products! I cannot find a parallel in any other industry. If farmers, for example, can bear the cost of transporting their produce to the market in an expensive way because of bad roads, the consumer, like our Clients, will pay for that cost in the end and transporters will probably earn money. But, unfortunately in the survey control issue there is no transporter to gain only wastefulness at the expense of our clients and the economy prevails.

ii.   Composition of the Survey Team

In this regard we need to look at the mixture of professional and non-professional staff (opportunity cost), or in other words, qualification and experience of the team on one hand and their quantity on the other. A rule of the thumb, in the division of labor is that each shall only be occupied with only that work which subordinates cannot and are not trained to do. A simple combination is one professional to at least three technicians and one technician to be in charge of at most seven assistants. These numbers may increase in a situation where some technicians and assistants are capable of playing a supervisory role to a maximum of double the stated numbers. Top-heavy teams are unproductive both technically and administratively and costly on labor. I would like to suggest the following seniority categories based on academic and professional qualifications as well as experience. In good paying environment the ration of the top and bottom levels in a team shall be greater than one to ten.

iii.   Market Cost of Plant and Equipment

The cost of renting out surveying equipment, like any other technological instrumentation, is a variable factor. Consideration is given to its landing cost, replacement value, devaluation rate and availability on the market. These factors suggest that the more modern the technology the higher would be the cost of its deployment in a project. The high cost of state-of-art technology is often profitably offset by its high productivity.

In many cases the hiring firm in Tanzania will insist upon sending own personnel with equipment who will have to be paid TAS 30,000/= on the average per day per person and thus nearly doubling the rental charge. Surveyors must be aware that quality assurance is not receptive to such imposition. But whether or not the Land Surveyor makes use of the escort, payment will

still have to be made. This amount enters the projects
as a direct overhead (technology cost) expenditure.

Chapter 6: TENDERS AND CONTRACTS

The Need for Finance

A business must be supplied with finance at any moment it is required. Where there is a regular inflow of receipts from sales and a regular outflow of payments for the expenses of the operation no serious problem arises, but in many cases a considerable time must elapse between expenditure

Sources of Finance

It is the purpose of financial institutions to assist in the financing of business during this interval. Business firms look to the capital market and the commercial banks to assist them. In the case of the state, the revenue comes in mainly in the fourth quarter of the financial year, so recourse has to be made to treasury bills to finance expenditure during the earlier part of the financial year.

In the public sector, one project is usually at the expense of another. A public sector project might induce the state to raise taxes. In this regard the project would be at the expense partly of private consumption and partly at the expense of private investment. A public sector project might cause the public to increase its borrowing from the public at the expense of private consumption and investment with an increase in the national debt. Public sector projects must be designed in such a way that one project will not affect total taxation of borrowing.

Financing Tenders and Contracts

Borrowing to finance production or project execution is a normal part of business affairs. A good business atmosphere requires, therefore, that finance institutions be available and ready to lend money and businesses be credit worthy. Finance related needs in tender and contracts include:-

• **Tender Bond**. This is an undertaking by a commercial bank jointly and severally, with their client who is submitting a tender bid (the tenderer) in which the bank provides surety in a given sum (about 1% of the tender sum) and for a specified number of days as a tender bond. The conditions of the bond are such that:-

1. If the tenderer withdraws his tender during the period of validity specified; or,

2. If the tenderer refuses to accept the correction of errors in his tender in accordance with the method provided in response to the instructions to the tender; or

3. If the tenderer having been notified of the acceptance of his tender during the period of tender validity, fails within the time stipulated to either

execute the contract agreement or to furnish the performance security bond; the bank will pay an amount of money equal to the value of the tender bond on first written demand.

•   **Performance Bond undertaking**. This is also an undertaking by a commercial bank to provide their client (the contractor) with a bond or guarantee in a sum stipulated in the conditions of contract (10% of the contract sum) for the due performance of the contract by the contractor. In other words this is a guarantee for the availability of money for the execution of the contract at inception stage.

•   **Insurance Undertaking**.   This   is   an undertaking by a reputable insurance company to provide insurance cover to a contractor on a project. Normally the business executing the project will obtain written guarantee from their bankers for a fee and if in default the money will be paid as promised and deducted from the business account with a premium.

•   **Letter of Credit.** To allow smooth payments of large amounts of money, that are not normal in daily banking operations, to the project executor by the client or employer. A commercial letter of credit contains instructions by the employer to its banker and the instructions are irrevocable as long as there is no breach of contract by the executor. The LC will have a period of validity.

Land Surveying Financial Issues

Land surveying projects are particularly in need of financing for, as stated earlier; finance must be available when required. A survey firm in Tanzania must therefore have reliable backing and not rely on contractual payments alone. Receipts from executed work are erratic for the following reasons:-

i.   The duration of execution cannot be precisely established as the execution of survey projects are greatly influenced by external weather conditions and the seasons,

ii.   The speed of work cannot be established early in the project until adequate reconnaissance has been undertaken and completed. Reconnaissance seeks to establish conditions with the datum, establish lines of communication, verify visibility and accessibility conditions and identify sources of logistics,

iii.   The completion of cadastral work, in particular, does not entirely depend on the executor but on the examination process. The duration of examination and approval is any ones guess. Clients in Cadastral surveying works would not pay until the final product has been delivered. On this score it has been advised (Dale, 1976, p185) that "There is no point in operating a professional licensing system if every survey has to be thoroughly checked; either the professional should be made to take fuller responsibility for his work or else the work should be given to technicians operating under Government supervision."

References

1. American Association of Consultants: Manuals for Members
2. World Bank. Standard Form of Contract

PART 2: SOME GEOMATICS PROJECTS UNDERTAKEN BY TOPO-CARTO CONSULTANTS LTD IN TANZANIA

Five projects are presented to illustrate the application of the preceding text. These were undertaken under the management and execution of the author who also wrote the reports being presented. The contracting company was Topo-Carto Consultants Limited of Dar es Salaam, Tanzania. They are of a wide variety covering estate divestiture, dredging, data audit in a route location for highway construction, re-titling for individualisation of title, and topographic site modelling for construction of a cement factory. All five projects had a different focus in project management and own issues to be addressed.

PROJECT No.1: DIVESTITURE OF SISAL ESTATES IN TANZANIA, 1997

This report has been prepared in response to client's letter of 19th July, 1999. The contract has provision for two reports namely Draft and Final Reports. The former was submitted to you in May last year and certified. Topo-Carto is now preparing the final report to be submitted upon completion of the Work. In addition we, on 20th January, 1999 submitted an interim report to you.

We therefore take it that the status report is to cover the period after 13th May, 1998 and the consultations we have had with you in that period on this project. May I first outline the critical activities in this project and the success or otherwise of their accomplishment.

OPERATIONAL STEPS

1. Signing of Contract: 18th September 1997.
2. Data search, field Reconnaissance and Search for Existing Beacons.
   - this item was comprehensively reported upon in the interim report of 20.01.98 and the draft report of 13.05.98.
3. Application for consent to subdivide the estates under title. This is a legal requirement as per

section 83 of the Land Regulation Act and quoted on our all offers all farms first applications for all Estates made on 19th November, 1997 was presumed last by Commission while searching for files. Our second application was lodged on 12th February 1998. Appropriate files could not be found.

- The schedule officer requested the office supervision to open new files on 5th April 1998. The application was complete for all 8 estates.

4. Consent of the Commissioner to subdivide the Estates. We are informed that the opening of the 8 temporary files were completed in October with Toronto and Mnazi (September) Gomba, Usambara and Ndugu (early October) Mwelya, Mkumbara and Mwenga (November) Consent was granted for the three groups respectively in September, November and December 1998. By Christmas 1998 Topo-Carto was ready to undertake the next legal step.

5. Application for Survey Instructions.
- The application were for subdivision of the land and excision part there of from the survey plan. The application were made in three batches dated on 27th November, 1998 and 17th December 1997 and submitted in mid-

December and early January personally to the Director of Surveys and Mapping. These were instantly approved and forwarded to schedule officers for formalization.

6. Granting of Survey Instructions.

   Although the Director order by Christmas, 1998 that Survey instructions be issued three were attended to on 4th and 7th January, 1997. The remaining were neglected and later presumed lost. We are yet to receive instructions in writing for the remaining five estates.

7. Execution of Surveys, computations and Survey Plans.

   - Topo-Carto took the contract signed with PSRC with all the seriousness it deserved and commenced services immediately upon signing even before the advance payment had been paid out to us. We went into field surveys at own risk even before receiving survey instructions to expedite the work. After a temporary demobilization between January and April due to El Nino storms were resumed and completed all field work for Client's certification on 13.05.98.

8. Submission of Surveys for Approval.

   It is a requirement that surveys can only be examined for approval by the Director for

Surveys and Mapping instructions had been issued by the Director.

We therefore had to wait till after the Directors' indication for no objection/to issue the same. We therefore submitted our surveys for approval by mid-January 1999.

9. Aproval of Surveys

    - Is awaited

10. Applications for new certificates for title.

    - This has been made in respect of Gomba and Mwenga sisal estates only whose certificates of title were revoked by the President on 4th April 1991. The circumstances surrounding the revocation were covered in both the Interim and draft reports.

    After long consultations with various schedule officers at the Commissioner's Office, I was then advised to forward the matter, with recommendations, for the Commissioners decision. My letter dated 28th January, 1999 addressed this subject. The reply look time to come through till June 1999 stating that new CT, be issued in the name of TSA. Subsequent to that I received on 13th July 1999 letter of offer for the two estates. CTs will be prepared as soon as the years land rent and pertinent are paid to the ministry.

11. Amendments to certificate of Title.

- The objective here is to effect changes which will rest ownership of the farms in the new investors at an agreed acreage which is smaller than that stated on the titles taken together for each estate.

All certificates of title (CT) can be categorised in one of three groups namely:

- CTs facing excision survey and part surrender of land.

- CTs which are free of encumbrance

- CTs in group 1 and 2 that are registered in names other than TSA.

Group 1 certificates require a surrender deed to be prepared and duly signed. In the event that the CT is registered in names other than TSA a deed of Transmission by operation of law will have to be included. The titles will then be transferred to the investors.

Group 2 certificates have to be transferred to the investors directly except those registered in names other than TSA. In the latter case a deed of Transmission by operation of a law will be prepared.

All documents prepared by Topo-Carto Consultants and signed by TSA Managing Director and attested by our lawyer were submitted to the Commissioner for Lands on 15th march, 1999. These were found erroneous in that PSRC and not TSA is the rightful signatory. The documents were withdrawn and returned to PSRC for action on 17th March, 1999. Again PSRC was not happy with the suggestion and were returned to the consultant. Topo-Carto then requested the Commissioner's staff to prepare deed of surrender (part surrender). The Regional Land Officer in Tanga supplied the deeds to Commissioner of Lands on 14th June 1999 for finalization. When ready Topo-Carto Consultants Ltd. will deliver these for signature of the Executive Chairman of PSRC. At the same time we hope that PSRC will have completed preparation of deeds of Transmission by operation of law and deeds of Deed plans will be prepared as soon as surveys are approved.

12. Final Report to Client

- The final report is a contractual requirement to be submitted and certified at the end of the assignment. The final report will be ready one week after the titling process has been finalized.

PROBLEMS, CHALLENGES & SOLUTIONS

The problems that have delayed the resurvey and titling work for estates on have been discussed several times with Client (PSRC) officials and one worth mentioning here:

1. Field disturbance and uncertainty.

    In this group we single out the El Nino rainstorms of 1997.98 and loss of Boundary beacons on estates. My team was forced to demobilize after the average working day came to two days a week due to the heavy down pour and roads became impassable, Further we took longer time than anticipated to re-establish the survey datum on estates as most beacons were presumed lost.

2. Inefficiency and Indifference at the Ministry of Lands and Human Settlement Development comprising both the Commissioner for Lands and Surveys and Mapping Division.

A   Commissioner for Lands:

1. Our original application for subdivisions of estates submitted on 19.11.98 were lost there. We were asked to resubmit the applications in Feb. 1998 yet files could not be found. Only

after 4½ months did the schedule officer seek the opening to temporary files. Yet is was only 19 months after applications were first submitted that the first batch was forwarded to the Commissioner for action and 13 months that the final consent was granted. (13 months delay).

2.	The officials were appraised of the revoked titles as early as September 1998. The suggestion to apply for new CTs was made in January 1999. Even after submission of applications for new CTs it took 5 months for decision to be made 1 month to write 2 letters of offer (6 months delay).

3.	Processing deeds of surrender (part surrender) The Commissioner while granting consent to undertake excision surveys instructed the Regional Administrative Secretary in Tanga to prepare surrender deeds for the her signature as early as September 1998. The Regional Land Officer only submitted these (also uncompleted to the Commissioner on 14th June 1999 (9 months delay).

Further the printing of the (affidavit) deeds have had problems since our request was made in April 1999. Typical problems were absence of officers on budget meetings, power failure and lack of crested paper. The last one remains critical.

B. SURVEYS AND MAPPING DIVISION

This is the department that grants survey instructions, a key procedure in cadastral surveys, examines and approves and cadastral surveys. Survey instructions could only be sought after receiving the Commissioners' consent stated earlier. The Director was most efficient, on these, to issue directives to schedule officers to deal with the matter. The same efficiency prevailed on 3 applications and stalled, presumably misallocated to date (6 months delay).   All surveys were submitted in January are yet to be approved (6 months delay).

4. LIASON WITH PSRC

Our requests made to PSRC officials were rarely followed up to successful conclusions.

A. We long awaited for a solution to mission certificates for Title such as those lodged with LART, mission in transfer from TSA and PSRC. The latter is still pending.

B. Who is to prepare transmissions and transfers and who is the rightful signatories. We had to find out the hard way and still not very sure as the relationship of PSRC to TSA is confusing.

C.	Names and addresses of investors could not be obtained through written communication. Officials we asked to confirm our list for example Southern Highland Estates is no Highland Estates is it?

D.	Issues on acreage in the case of conflicting information.

We have always been of the opinion that consultants have to be supervised by a person knowledgeable in cadastral surveys, a licensed surveyor, with experience in Project Management. Regular (bi-weekly or monthly) meetings are necessary to follow through the work methodology and time schedule, among other things.

## 5.	FINANCE

The project was designed to last a maximum of 7 months and resources were solicited from donors to accomplish the task within that time frame. In other words the funding covered the project expenses and, say 15% profit to the Consultant. Delays attributed to causes beyond the consultants reasonable control as described above have made the financing unmanageable and project execution less efficient. The client is requested to seriously consider granting a topping-up as requested by all consultants by letter Ref. No. PTB/22/99/01 of 2nd February 1999. The approximately 15% increase accepted by client this month will definitely help but not suffice.

signed
Dr. F. N. Lugoe
Lead Consultant
Topo-Carto Consultants Ltd.

FINAL REPORT ON ESTATE DIVERSTITURE

Background

On the19^{tth} September 1997, Topo-Carto Consultants Limited (T-CC Ltd) signed an agreement with the Presidential Parastatal Sector Reform Commission (PSRC). The agreement called for the re-survey and titling of eight estates of the Tanzania Sisal Authority (TSA), comprising zone D. This group of estates is made up of Mwelya (and Usambara), Toronto, Mkumbara, Gomba (and Mwenga), Ndungu and Mnazi sisal estates.

The Consultant was to undertake this task in order to assist the Client achieve two objectives. These are; to obtain approved survey plans and subsequently obtain certificates of title in the name of the buyer for the land held by TSA on each estate after excising the land that has been encumbered.

The scope of work in the agreement required T-CC Ltd to firstly, do all necessary preparatory work and subsequently carry out cadastral surveys for the fixation and approval of <u>new</u> boundaries that will allow part surrender of agreed acreage of land from each estate. Secondly, T-CC Ltd would <u>process</u> amendments to existing Certificates of Title or obtain new offers, as the case may be, that will eventually vest ownership in the names of the prospective investors.

Adjudication

The purpose of this exercise is to ascertain that the land holding of each estate is as given in the terms of reference to the contractual agreement and, conversely, that the acreage given in the scope of work is exhaustive. In other words, adjudication ensures that, upon completion of the re-survey and titling process, all land owned by each estate will have been apportioned either to the new investor or the local Government in accordance with the agreement between the two parties.

In this project the adjudication process involved the following operations:

- obtaining the certificates of title;

- acquisition of acreage information from the certificates of title (available either at TSA headquarters or the zone registry of the Commissioner of Lands in Moshi);

Table1:   Status of Zone D Estates as per Adjudication.

Estate	C. T. No.	Area (Ha.)	Encumbrance	Remarks	Action
Mwelya	6138	1650	yes	CT presumed lost	Excise 235 Ha,
Usambara	3787	415	Yes	CT presumed lost	Excise 90 Ha,
	12890	14.2	None	ditto	
	15920	562	None	ditto	
Gomba	4527	2749	yes	CT revoked	Obtain Offer to Land
Mwenga	16397	1692	yes	CT revoked	Obtain Offer to Land
Toronto	8317	2116	Yes	CT available	Excise 400 Ha
	4644	2428	None	Ditto	Process CT amendment
	4645	1008	None	Ditto	Process CT amendment
	8316	7.7	None	Ditto	Ditto
	4727	930	None	Ditto	Ditto
					Ditto
Mkumbara	5226	1127	Yes	LART Mortgage	Excise 522 Ha
	8059	300	Yes	Ditto	Recover, process

	15600	830	Yes	Ditto	amendments Excise 8 Ha, recover and process amendments . Excise 47 Ha, recover and process amendments
Ndungu	7748	663	Yes	Certified copy of CT only	Excise 48.5 Ha, process amendments to the CT
	15427	623	Yes	CT presumed lost	Excise 76.5 Ha,
Mnazi	11247	988	Yes	CT available	Excise 70 Ha
	17146	481	None	CT available	Amended by investor
	4144	227	None	CT available	Amended by investor

- the physical identification and location of land;

- verifying the acreage to be surrendered and to ascertain its correctness in terms of the agreement between TSA and Local Government;

- establishing the land ownership or title transfer scenario in terms of the law and

Inspect the agreed lines of demarcation that have been delineated on the ground by the parties involved in the part surrender process.

It was established, by the consultant and confirmed by TSA management, that a delineation process and physical hand over of land to villagers had taken place by the parties. Following the hand-over cum take-over process the villagers proceeded with land development in the areas granted to them. The estate management also consolidated their farm development. T-CC Ltd consulted with the client on the issue as the delineated land was likely to differ in acreage with that agreed at round table talks. It was then accepted that physical delineation was an act of agreement of the size of land envisaged in the excision and the delineation must therefore take precedent over acreage numbers during the surveys.

The information gathered during the adjudication process has been summarised in table1 above and to a large extent defined the workload of the consultant both in the office and in the field. Further, the client was appraised of the possibility of reduction in acreage where water front areas were to be removed from titles as per instructions. It was then agreed that the area remaining for sale should not be reduced. An overriding assumption being that the water front areas will be for the benefit of the villagers. The area for sale was also to be maintained as closely as possible since it has been entered into the sale agreement in accordance with the tendering documents.

Field Surveys

After due preparations, T-CC Ltd dispatched its field team to commence work on the estates on 12th October 1997. The field operations included

- the search and recovery of boundary beacons,
- establishing survey control on the estates,
- datum checks,
- field computations,
- selection of sites for excision beacons,
- construction and erection of beacons,
- excision and subdivision surveys,
- computations and plan drawing,
- signing of boundary beacons with estate management.

Office work was even more involving. The various activities carried out as part of office work included:

- data search,
- pre-computations,
- search for consent of the Commissioner of Lands to divide registered land,
- search for survey instructions,
- survey adjustments and computations,
- preparing the data base,
- preparation of computer sketches,
- computer plotting of survey plans,
- data base verification,
- verification of excision acreage,
- vetting survey data,
- writing survey reports,
- submission of survey reports for approval, and registration of survey plans,
- writing contract reports to client.
- Writing contract reports to client.

Title Search

Titling work commenced shortly upon signing of the agreement by obtaining, where possible, copies of certificates of title and other relevant information regarding land ownership and associated encroachments on each estate. In adjudication, the Consultant had the responsibility of sorting out the certificates of title and identifying those likely to be affected by part surrender. Encroachments were finally identified both on the ground and on the map unambiguously.

In accordance with section 83 of the Land Registration Act (Cap 334 – sup. 65), Form LR 30 (applications for division of registered land) were completed in respect of each title that was a subject of land surrender. These were submitted to the Commissioner of Lands for approval on 19th November 1997. Unfortunately, the applications were later reported missing. T-CC Ltd therefore submitted new applications on the12th February 1998. All applications were accepted and endorsed by 16th December 1998, paving way for the Consultant to apply for survey instructions before Christmas of 1998. Details of this process have been provided in consultant's status report of July 1999.

On accepting that the estates be divided in terms of the Land Registration Act, the Commissioner also issued a directive to the Regional Administrative Secretaries (RAS) regarding the part surrender process. In that directive, (references LD/1353/8, LD32967/243/STM, LD/1789/10, LD/26591/8/TNN, LD/1684/9/TNN, LD/1352/3/TNN) the RAS would prepare the deeds of part surrender and submit these for the Commissioner's signature. The signed deeds would then be passed on to the Registrar of Titles to complete the part surrender process. Kindly consult the appendix for copies of the directives for each estate.

The consultant was also advised by the Commissioner's staff that title transfer deeds would be prepared by the liquidator (PSRC) of the estates. These would be in PSRC's seal and signature. As a result of this development, all deeds prepared by the consultant's lawyer were not used.

The case for Gomba and Mwenga estates was different from other estates in zone D in that the two had no Certificates of Title at all. The original titles were revoked in 1991. The consultant presented a letter requesting the Commissioner to issue new certificates in this regard. The procedure involves obtaining offers first to be followed by the certificates upon payment of land rent.

End Results

Summary of information on land parcels in zone "D" estates that <u>were affected</u> by encroachments and hence land surrender upon completion of the resurvey and titling project is tabulated hereunder.

Table2:   Land Tenure Data.

Name of Estate	L.O No	L.D No	C.T No	Pre-sale Gross Area (Ha)	Net Surrender Area (Ha)	Post-sale Net Area (Ha)
1	2	3	4	5	6	7
Mwelya	-	26591	6138	1650	238	1412
Usambara	9280	1684	3787	991.2	91	900.2
Gomba	22643	1789	4527	2749	1044	1705
Mwenga	564, 565 573, 574	1789	16397	1692	762	930
Toronto	10981	28349	8317	6489.7	401	6088.7
Mkumbara	558	2089	5226 8059 15600	1127 300 830	47 8 522	1080 292 308
Ndungu	559	1353	7748 15427	663 623	48.5 76.5	614.5 546.5
Mnazi	14116	3296	112477	1697	117	1580

Documentation and Services

The Following documents have been prepared in fulfilment of the objectives of this project as provided in the scope of work of the Consultant.

i.   Survey plans, with pertinent survey data and information, dully approved and registered by the Director of Surveys and Mapping. Information on the registered plans in provided in table 3.

ii.   New offers for Certificates of Title in the name of TSA for Gomba and Mwenga sisal estates. These are the estates for which the Certificates of Title were revoked by the President in 1991.

iii.   Deeds of Part Surrender on the remaining estates as part of the processing required in the scope of work. To complete the process of title transfer, the buyer will also require duly attested Forms LR 5 for "Transfer of a Right of Occupancy" and Form LR 25 "Application for Registration of a Transmission by Operation of Law".

iv.   Consultants report to Client. A total of four reports to the Client have been prepared namely, interim, draft status and final reports. The interim report was submitted to the Client on 20th January 1998. The draft report was certified by the client on

13th May 1998. The status report was submitted on 29th July 1999. This is the final and conclusive report on this project.

v. Advise on various aspects of the project provided to client or others on request of the client.

Timing and Scheduling

The total time accorded to the contract, including report writing, was seven months. However, this project required time extensions because of a number of reasons.

- Firstly, rain storms and floods in the project area prevented normal progress of work. The Client was appraised of the adverse effect of the heavy El Nino rains on the field operations as early as 24th November 1997. A 2-month extension of time was then granted. As it turned out the stoppage due to the El Nino Storms were more erratic and longer requiring a time extension of three (3) months and four mobilisations and demobilisations.

- Secondly, pursuant to the general conditions of contract (appendix C) the client issued a variation order to the Consultant with regard to Gomba and

Mwenga estates on 5th March 1998 requiring an extension of time of 3-weeks time with pay.

- Thirdly, it was agreed with the client that the delay by the Commissioner for Lands in processing applications for the grant of consent to divide land under registered titles was beyond the Consultant's control. The causes were cited as frequent reorganisation in the Lands Department, personnel absenteeism, retrenchments and loss of files. All documentation has finally been done using temporary files. The delays costed the project twelve (12) additional months. There were also extended unreasonable delays in signing the surrender deeds by the Commissioner of Lands lasting at least another twelve (12) months.

- Finally, There were extended unreasonable delays in approving the survey plans. The process that normally would take one month lasted eleven (11) months

The delays are detailed in the status report of July 1999.

Payments

The signed agreement provided for an advance payment of 20% of the contract sum and two further equal payments of 40%. The cheque for the advance payment was received by T-CC on 14th October 1997. Payment for the first phase of the work under this agreement was certified on 13th May, 1998 and payment received on 10th June, 1998. Payment in respect of the variation order on Gomba/Mwenga was made on 10th September 1999.

An additional sum to the contract sum was negotiated by the consultant to cater for delays in project implementation due to forces beyond the consultant's control as discussed in the previous paragraph. The amount granted, following this application, constituted about 15% of the contract sum.

CADASTRAL SURVEYS

Common to all estates in the project was the operation of the re-establishment of the survey datum in which the survey plans are registered. Such work commenced with a search for enough beacons along the boundaries necessary to anchor a control network free of datum defects. The beacons were necessary in this regard because the beacons are the only established survey markers that have been co-ordinated in the survey system of the registered plans.

The search for beacons proved to be an arduous task. Most of the beacons erected in easily accessible areas were, for various reasons, missing. The inaccessible areas required bush clearing, in addition to hours of walking to the 'would be' locations. This exercise also depended on visual inspection that relied heavily on the local population some of whom last visited the place many years before the vegetation overgrowth that covered the area during project execution. Search efforts were also prevented by the heavily pouring rains.

It took the team between 1 and 4 weeks, on the average, to re-establish the datum on an estate. On most estates the found beacons were not intervisible. The task of datum checks, in this case, was combined with that of establishing and extending control on the estate for better results.

Methodology

Methods chosen in control surveys were triangulation (including triangle solutions), traversing and radiation incorporating distance checks. The equipment employed in the measurements were - Sokkisha total stations, SET3B and SET2C, Wild distomat, DI3S mounted on Wild T2 theodolite and AGA geodimeter AGA-200 mounted on Wild T2 theodolite all backed with sets of CB radios for communication in the field. Initially, triangulation was the favourable method but as shrubs and grass grew with the increasing rains to shoulder height, the method demanded more bush clearing than others. This fact, with growing uncertainty in point location in the field, meant that the method was to be disbanded in favour of traversing. Radiation with distance checks was used where datum beacons were located on high ground and the new points were intervisible.

The processing of observations and computation of co-ordinates and their derivatives has been accomplished on all estates. These were the basis for the computation and verification of final values regarding acreage and plotting the survey plans. A co ordinate list of all beacons demarcating the estate after the excision surveys and copies of survey plans are provided with this report.

Survey Report

A survey report was prepared in close conformity with the computing instructions and SMD technical circulars. Eight such reports were submitted to the Director of Surveys and Mapping for approval and record keeping. Upon approval, the DSM had therefore approved the survey plans for each of the eight estates. Copies of survey plans were submitted to the Regional Administrative Secretary (RAS) in Tanga to assist in preparing surrender deeds. The relevant parts of survey reports submitted to the Director of Surveys and Mapping are given in the appendix to this report. All surveys were carried out in accordance with survey regulations as stipulated in the Land Survey Act (Cap. 390 sup. 57).

Table 3.   Survey   Plan   Registration Information.

Estate	Farm Numbers		Plan Number	Registered Plan Number	Registration Date
	Old	**New**			
Mwelya	48	680	E^4 163/1	32298	27.10.99
Usambara	50	679	E^4 87/3	32441	30.11.99
Gomba	46	682	E^4 122/3	32314	09.11.99
Mwenga	43, 44	683	E^4 122/4	32446	30.11.99
Toronto	-	675	$E^4$110/1	32313	09.11.99
Mkumbara	133	678	E^4 80/1	32297	27.10.99
Mnazi	292a, 292b	674	E^4 228/1	32315	09.11.99
Ndungu	186	1032	E^4 78/1	32300	28.10.99

Table 3 above provides information on each estate as per approved and registered survey plans. This information pertains to those parts of the estate that have been subjected to excision and part surrender of land. The information remains unaltered where excision and part surrender was not deemed necessary. The plan numbers are handy in, among others, obtaining copies of the approved survey plan from Government records kept in the 'records office' of the Division of Surveys and Mapping. The office is currently housed on the 4th floor of Ardhi House in Dar Es Salaam.

The farm numbers are unique for the farms and may be displayed by the farm owners on billboards as deemed fit. Where new numbers have been issued, the old ones ceased to be used as of the date of registration.

TITLING WORK

The search for title was accomplished in close conformity with the existing legal provisions based on technical requirements of cadastral surveying. The terms and conditions imposed on all certificates of title call for the occupier to, among other things, satisfy the requirements of section 83 of the Land Registration Act . It is explicitly stated in some of the certificates that "The occupier shall not subdivide the land or assign, sublet or otherwise dispose of it or of any part of it without the previous written consent of the Commissioner for Lands". Further, as a matter of procedure, the Director of Surveys and Mapping could only issue survey instructions when the consent of the Commissioner had been produced as evidence of permission to undertake excision surveys on land under title.

It was therefore necessary for the Consultant to apply for the Commissioner's consent as an initial step in this undertaking. It was also necessary to await the issuing of survey instructions before seeking approval of cadastral survey work. There was much unnecessary delay (11 to 13 months) in obtaining the needed consent. The survey work had to wait therefore eight (8) months to be submitted to DSM for approval. As a consequence, the registered survey plans upon which all amendments to existing titles are based were only available in February 1999. At that time the search for amendments to the Certificates of Title was initiated by preparing documentation for the part surrender of agreed acreage of land and transfer of title. Offers for new Certificates of Title have been processed only for Gomba and Mwenga estates whose titles were revoked in 1991.

Methodology

Two approaches to the provision of title were identified and used in this project. A search for amendments, to be made on the certificates of title, was undertaken for all titles relieved of encroachments and which were a subject of part surrender of land on the various estates. The table below provides information on this group of certificates.

In Table 4 below, 83 Ha were excised from the river frontage at Gomba and returned to the President and the 48 Ha error in the C.T on Mnazi estate are also included as excised and therefore surrendered land.

Table 4:   Data on Titles with Land to Surrender.

Estate	L.O No.	L.D No.	C.T No.	Gross Acreage (Ha).	Surrender acreage (Ha)	Remainder Acreage (Ha)
2	3	4	5	6	7	8
Mwelya	-	26591	6138	1650	238	1412
Usambara	9280	1684	3787	415	91	324
Gomba	22643	1789	4527	2749	1044	1705
Mwenga	564, 565 573, 574	1789	16397	1692	762	930
Toronto	10981	28349	8317	2116	401	1715
Mkumbara	558	2089	5226 8059 15600	1127 300 830	5228 47	308 292 1080
Ndungu	559	1353	7748 15427	663 623	48.5 76.5	614.5 546.5
Mnazi	14116	32967	11247	988	117	871

Table 5: Data on Titles **Not** Affected by Land Surrender

Estate	LD No.	LO No.	CT No.	Acreage Per CT (Ha)	Total Area (Ha)
2	3	4	5	6	7
Usambara	1684	9280	12890		576.2
		9280	15920		
Toronto	28349	9260	4644	2428	4373.7
		8736	4645	1008	
		8784	8316	7.7	
		572	4727	930	

Table 5 above documents the information for certificates that are not a subject of land surrender but direct title transfer to the investor. Title transfer in respect of certificates of title numbers 4144 and 17146 of Mnazi sisal estate were completed early in 1998.

ESTATE SPECIFICS

Toronto Sisal Estate

Toronto estate had a total of 15871 acres (6423 ha) of land on five certificates of title before excision was effected. The certificates are marked with numbers - 4644 of January, 1938, 4645 of April, 1938 4727 of May, 1939, 3816 and 8317 of 1952, for 6001, 2492, 2298, 19 and 5227 acres respectively. The ownership of all five estates had been transferred into the name of TSA on the 2nd July, 1991.

Only one certificate of title on this estate was affected with excision in this project, namely title number 8317. A total of 401 Ha were excised for surrender to the President of the United Republic of Tanzania.

Mnazi Sisal Estate

There were two issues of land surrender encountered on Mnazi estate.

The first was that the acreage of 2442 acres stated on CT No. 11247included 118 acres that was a subject of "land exchange between Native Authority and Mnazi Estates Limited at the time that the CT was issued. Existing survey plan E4 228 shows that an area of equal acreage was later added to the estate but was overlooked in the statement of acreage in the Certificate. On the ground, the land ceded by Mnazi Estates Limited of 118 is used by the natives and not the estate. A rectification has been made with a request for the Registrar of Titles to remove the acreage from the statement of acreage on the Certificate of Title.

The second was the land of 170 acres (69 Ha.) for part surrender that was accepted by the new owner, M/S Le Marsh Enterprises against the 297 stated in the terms of reference. The issue regarding the difference was agreed upon with PSRC and as detailed in consultant's letter ref:TCC/S197-31/FNL of 27th November, 1998 to Executive Chairman, PPSRC.

All in all a total of 288 acres (117 Ha.) of land were surrendered from CT number 11247. Title transfer for the other two certificates was completed early in 1998 and required no further action by the Consultant. The owner will have to be prevailed upon to ensure that the ceded acreage is documented in the certificate of title number 11247.

Ndungu Sisal Estate

The Estate has two titles registered under numbers 11527 and 7748. The land earmarked for part surrender was excised from both certificates of title.

Mwelya Sisal Estate

Excision has affected the single Certificate of Title for Mwelya Sisal Estate. All documentation has been completed.

Gomba and Mwenga Sisal Estates

The Certificates of Title for Gomba (number 4527) and Mwenga (number 16397) were revoked by the President of the United Republic of Tanzania on 4th April 1991. By this action, Banji Laxman's claimed ownership by power of sale from NBC had therefore been declared null and void. The revocation caused the ownership to revert to TSA by acquisition and subject to section 71 of the Land Registration Act.

The estate at the time of this project was owned by TSA by operation of a law but TSA had no certificate of title to their land. The Consultant was advised by the Land Officers to present the case and await decision to grant

TSA an offer followed by a title and subsequent transfer to the investor. The letters of offer for the two estates were received on 13th July 1999 in the name of TSA. The Certificates will be for a term of 99 years. The Commissioner of Lands rejected the consultant's application to make the offer in the name of the investor as "TSA would have nothing to sell".

Usambara Sisal Estate

The land parcel known as Usambara sisal estate has three certificates of title registered in the name of TSA. CT numbers 12890 and 15920 had no encumbrances and were therefore not a subject of land surrender. Land registered under CT number 3787 was duly excised as required.

Mkumbara Sisal Estate

The land parcel known as Mkumbara sisal estate is provided with three certificates of title (numbers 15600, 8059 and 5226). All three certificates of title had been under mortgage and later transferred to the Loans and Amenities Realisation Trust (LART). After quite an effort by the Consultant, the documents were only returned to PSRC on 7th March 1999 to enable the titling process on this estate to proceed. All three certificates had encumbrances and were resurveyed.

SUMMARY

The excision was carried out in strict compliance with the instructions of the Client and survey practice. In almost all cases there was a discrepancy between excision acreage provided in the terms of reference and the agreed physical lines of excision. The latter were followed since they were delineated on the ground. Subsequently, settlements and other developments had been undertaken by the recipients following these lines. Table6 below summarises the land tenure scenario on completion of the project.

Table 6: The New Land Tenure Scenario on the Estates.

Estate	District	CTs	Acreage (Ha.)	Investor	Investor's Address
Mwelya	Korogwe	1	1412	MGC Limited	P. O. Box 277, Tanga, Tanzania
Usambara	Korogwe	3	900.2	MGC Limited	P. O. Box 277, Tanga, Tanzania
Gomba	Korogwe	1	1705	Gomba Agricultural Industries Ltd	P. O. Box 423, Dar Es Salaam, Tanzania
Mwenga	Korogwe	1	930	Gomba Agricultural	P. O. Box 423, Dar Es Salaam,

				Industries Ltd	Tanzania
Toronto	Korogwe	5	6088.7	Highland Estates Limited	P. O. Box 1406 Iringa, Tanzania
Mkumbara	Korogwe	3	1680	D. D. Ruhinda & Company Limited	P. O. Box 1987 Tanga, Tanzania
Ndungu	Same	2	1161	L.M. Investments Limited	P. O. Box Tanga, Tanzania
Mnazi	Lushoto	3	1579.5	Le-Marsh Enterprises Limited	P. O. Box 69731, Nairobi, Kenya

CERTIFICATE

This report has been accepted by management of Topo-Carto Consultants Limited as a final report in terms of the agreement with The Presidential Parastatal Sector Reform Commission dated 18th September 1997.

for and on behalf of

Topo-Carto Consultants Limited.

…………………..

Dr Furaha Ngeregere Lugoe

Managing Director.

………. December 1999

PROJECT NO. 2: ENGINEERING SURVEYING AND THE DREDGING OF THE DAR ES SALAAM HARBOUR, 1997/98.

The key participants in the project are: THA-Tanzania Harbours Authority, now Tanzania Ports Authority, as Client, HAM-Hollandsche Aannemming Maatschappij of Holland as Contractor and Evers Consult of Australia in association with Topo-Carto Consultants of Dar Es Salaam as The Engineer. Project was accomplished under FIDIC-III terms and conditions of contract commencing on 26th May 1997 for an execution period of 39 weeks. Project was financed to the tune of USD 23.3 million and funded 60% by The Royal Netherlands Government (ORET programme) and 40% by THA. Material used in preparation of this report in addition to personal involvement and experience of the author, include:

- The 1993 tender documents (vol. 1-5) for the Entrance channel improvements dredging works drawn by Scott Bertlin and updated in 1996 by Evers Consult in association with Topo-Carto consultants.
- The 1994 report (vol.5) on the Dredging of the Dar Es Salaam submitted to THA by SLI Consultants of Canada.
- The 1996 tender evaluation report for the Dar Es Salaam port improvements dredging works

by Evers Consult in association with Topo-Carto Consultants Limited.

- The 1997 Contract Agreement (vol.1 & 2) between THA and HAM for the Dar Es Salaam port entrance channel improvements – dredging works.
- Monthly progress (1997 – 1998) reports of the Engineer to the Client on the Dredging of the Dar Es Salaam port.

1. INTRODUCTION.

The port that gave our capital city its name is an old one. The marine vessels that called in our ports establishing commerce with its hinterland several centuries ago have undergone a renaissance. Several generations of ships have come and gone and, until recently, the state-of-the-art vessels could no longer praise our port to be the ''haven of peace'' that it used to be. Navigation into the port for these large ships had for some time been difficult due to the many bends and the shallow controlling depth of the narrow entrance channel. In the early 1970s the custodians of the port namely, the Tanzania Harbours Authority (THA) commissioned a number of studies aimed at modernisation in the port. One of the proposals in the modernisation process was to dredge the harbour in order to give large ocean lines unrestricted entry into the port.

By the term dredging one understands the civil and surveying engineering activities undertaken in or under water and associated with the removal or transfer of solid material. Dredging therefore involves cuts, fills and slope design, slope protection and transportation of material in the water. It is commissioned in waters that cannot safely be navigated by marine vehicles particularly in rivers, and harbours. Dredging can be undertaken for purposes other than navigation such as removing debris and silt in lakes and reservoirs aimed at increasing both depth and storage capacity.

The uniqueness of dredging lies in the location of the site in or under the water. The location itself determines the methodology of dealing with under water topography, earthworks and slopes. It in turn calls for unique equipment for both the marine works and surveys. In this regard, dredging works rely entirely on floating equipment or ones aboard floating vessels.

The site also calls for a substantially different approach to dredging project management compared with other civil and survey works. Here, more reliance is placed on survey displays provided by the survey unit, as site inspection is mostly, limited to the onshore activities and supervision of the surveys. The surveys unit produces bathymetric charts showing the progress of the works on a daily basis. The daily volume of work done is also determined by computations made using survey data and displays.

It must be mentioned here that dredging is not an environmentally friendly activity. It causes great disturbance to marine life particularly, coral reefs and fish breeding. Environmental disturbance can only be brought under control using good environmental management practices that include control of the disposal activities of dredged material.

This paper gives an overview of the 1997-98 dredging of the port of Dar Es Salaam. It then discusses the land and hydrographical surveys undertaken in context of dredging with particular attention being drawn to chart datum issues and solutions.

DREDGING

2. PROJECT EXPECTATIONS.

For a long while prior to dredging, the conditions and characteristics of the entrance channel of the Dar Es Salaam port had been considered to be below international standards making the port less competitive with neighbouring ports. For example, before the recent dredging was completed, navigation in and out of the port at night was not permitted due to lack of reliable navigation aids and the many bends in the channel. Secondly, the controlling depth of 7.4 m along the channel allowed ships with a maximum draught of 6.7 m only at low tide, which limited the type of ships using the port. Thirdly, 'pilotage rules' for the port of Dar Es Salaam required ships that were longer than 122 m to enter and leave the harbour by *'stemming the current'* when the tidal range was greater than 2.9 meters. This range is far below the tidal range for the port, which can be as high as 4.2 m.

All in all, delays of up to 22 hours had been experienced in the port before this project was completed due to a combination of these adverse factors. Table 1 below show the characteristics of the port prior to and after dredging. Dredging has therefore enabled he navigation into the port of ships with twice the length allowed earlier and increased the draught to 10m as is common in ports of an international standard.

Table 1. Controlling Factors of the Entrance Channel.

Characteristic	Pre-dredging	Post-dredging
Channel depth	7.4 m BCD*	10.7 m BCD
Channel width	80 m	140 m
Time restriction	Daylight	None
Other restriction	Stem the tide	None
Max. current velocity	2.9 m/sec	2.0 m/sec
Max. vessel length	122 m	234 m
Max. vessel draught	6.7 at LLW	9.5 m at LLW

*BCD – below chart datum

Ships calling into harbours need assurances of safety, which, in addition to channel characteristics, include availability of safe leading lights and other navigation aids. Part of the dredging exercise was to place and precisely locate all navigation aids with respect to the channel centreline. Site, speed and accuracy conditions required the application of the latest survey technology in this activity as well.

3. PROJECT EXECUTION

The dredging works were carried out using mainly the trailer hopper suction dredger (Geopotes IX), cutter suction dredger (Sleidrecht 35), backhoe dredger (Hippopotes) and a number of survey and tug bats. The works in the project were divided in two main sections during execution, namely the entrance channel and harbour basin dredging. These were further divided into operational sub-sections as indicted in Figure 1 namely the outer shoal, middle shoal, centre shoal, Ras Makabe, Tafico shoal, east bank, Kurasini oil jetty Gerezani, west ferry point, east ferry point, berths 1-4, berths 506 and berths 7-11 each identified with specific dredging characteristics.

3.1 The Entrance Channel.

The approach from the sea over the outer shoal follows a straight line. Dredging was done to a minimum and included the removal of obstacles such as ship and navigation aid wrecks. The middle shoal constituted a complex curve in geometry, which had to be straightened. Both the middle and centre shoals were deepened. The channel was made to curve towards the berths for smoother enchorage of ships. The material dredged in the outer and middle shoals was of soft material and coral Most of the material was unusable and slope material for both west and east ferry points was therefore imported from granite quarries. Old navigation aids were removed as were wrecks and obstacles. New navigation aids were installed to mark the limits of the new channel.

3.2 The Inner Harbour.

Depths in the inner harbour were shallow and variable before dredging as it followed a natural hydro-geological formation. Dredging in this area involved deepening the sea bottom at the existing berths. Further work included the dredging of Ras Makabe, centre shoal, Tafico berths, east bank, Kurasini oil jetty and at the coastal jetties.

4 QUANTITIES ANS DEPTHS DREDGED.

Although dredging is primarily an offshore activity, it often involves excavations and reclamation on land as it happened in this project.

4.1 Offshore Works.

The dredged quantities and depths dredged in the various areas of the project are summarized In Table 2 below. Quantities were computed by the Contractor's survey team and verified by the Engineer's surveyors using Geocomp software version 5. A grid database method was used. Cells measuring 2.5 m square were created. A digital terrain mode (DTM) surface from the design and either a bathymetric chart was interpolated on to that grid. The computed volume would be a difference of the two.

Table 2. Measurement of the work done.

Dredging Area	Quantity in 1000 cub. m	Dredging Depth BCD	Remarks
Outer Shoal	537	10.90	
Middle Shoal and EFP	691	10.70	
Centre Shoal	205	10.70	
Ras Makabe	90	10.70	
Gerezani	58	10.70	
Tafico Shoal	105	4.80	
Lighter Quays	20	10.00	
Berths 1 to 11	285	10.00	
Berth 7 to 11		12.20	40 m wide
East Bank	55	10.50	
KOJ Min Berth	20	13.50	
KOJ	15	7.50	

Coastal Berth			
Total- excluding over-dredge	2429		

4.2 On Shore Works.

The widening of the channel between Magogoni and Kigamboni required a combination of on shore and off shore works. In so doing about 100 m of west ferry point had to be removed in a 1:4 slope. The slopes were protected with imported granite rock. A total of 348, 000 cub m. of material was removed from the west ferry point and dredged to a depth of 10.70m about a quarter of the dredged material constituted dry excavation. Part of the dry excavation involved the cutting of a coral boulder that had not been foreseen in the project, with severe damages to the dredger. Cutting of the boulder was a preferred option as it was too late in the project to seek blasting permission and import the explosives for the work.

Dry excavation involved relocating the existing ferry rump and market. On shore works included the construction of a temporary ferry rump at the end of Magogoni Street and demolition of existing structures. Further, the site was cleared, dry excavation done, and an area reclaimed. The site for the new ferry rump was prepared by building a perimeter bund that was filled with excavated material and protected with coral, spoil fill, rubble and crushed concrete from demolishing works.

5 ENVERONMENTAL PRECAUTIONS.

The speed and direction of residual ocean currents were assessed in considering he potential spreading of the sediment plume. In studying the available sea charts and conducting interviews with pilots, a likelihood of the occurrence of residual currents with speed equal to 1 to 3 knots in the NNW directions in the months of March to November was established. It was also established that for the rest of the year the residual current could be considered negligible for the purpose of this project.

Consequently, two deep dumping sites that would prevent re-suspension and keep the coral reefs safe throughout the dredging period were selected. The closest site was located at about 4km from the berths and the other at about twice that distance. Disposal during periods of very rough conditions was disallowed. As a further precaution, a tightly controlled disposal programme was designed for the contractor to follow in order to minimize diffusion. Along with these effective measures, it was made a requirement at tender stage that the hopper dredger to be used for disposal be among the largest, also to minimize diffusion. In order words, dump barges would not be allowed on the project.

During project execution, all dredged debris was disposed at as selected site about 4 km at sea to a depth of approximately 192m following a tightly controlled programme. This measure was instituted in order to avoid siltation to the coral reefs in the vicinity of the dumping site above acceptable level. Adverse effect was determined as a disposition of sedimentation on any of the surrounding reefs in excess of 1 mg/sq-day. In this regard, the environmental monitoring team installed silt-measuring receptacles on the reefs, which were collected every fortnight. The tubes were taken to the sedimentology laboratory of the geology effect of the Msimbazi River on the reefs to avoid misinterpretation of the results. The results confirmed the selection of the dumping site and no alternative site was therefore necessary.

ENGINEERING SURVERYS

6. SURVEY SPECIFICATIONS.

Survey work, in this project, was guided by several statements in the specification that required the contractor's compliance to the satisfaction of the project engineer. The Contractor had to provide a detailed method statement for all surveys undertaken and could only commence surveying after the engineer had approved the statement.

6.2. Scope of Surveys (clause 611)

The Contractor shall perform all necessary survey work required to be executed by him as directed by the engineer. For this purpose he shall furnish and employ all necessary survey work and employ all the personnel, services, equipment and supplies needed to perform the surveys and all the incidental work required for;

- The setting out, including the marking of all points for the works.
- The registration of water levels
- The pre-dredge survey and record drawings.
- The intermediate progress surveys and record drawings.
- The post-dredge surveys and record drawings.

6.3. The Setting-Out Grid (Clause 109).

The setting out grid for the works shall be the local Dar Es Salaam grid which is shown in the drawings. Appendix A1 lists the co-ordinates of fixes points, which were used in March/April 1973 by Wimpey Laboratories Ltd.

6.4. Tide Levels (Clauses 107 and 108).

The port of Dar Es Salaam is listed in Volume 2 of the British Admiralty tide tables. Tide levels are as follows (see chart in Figure 2):

- Highest Astronomical Tide (HAT)
  +3.7m
- Mean High Water Spring (MHWS) Tide
  + 3.2m
- Mean High Water Neap (MHWN) Tide
  +2.1m
- Mean Sea Level (MSL)
  +1.5m
- Mean Low Water Neap (MLWN) Tide
  +1.0m
- Mean Low Water Spring (MLWS) Tide
  -1.0m
- Lowest Astronomical Tide (LAT)
  -0.5m

The Contractor shall install an automatic tide recorder in the existing wet well within four weeks of the date of commencement of the works and shall arrange for the first thirty one days of data to be processed, tidal constants determined and the relationship between Admiralty chart datum and Harbours Datum established. The Contractor shall then establish temporary benchmarks relative to Admiralty Chart Datum at all his work sites.

6.5. Surface Levels (Clause 404).

Before commencing excavating or filling the Contractor's agent and Engineer's Representative shall jointly survey the area where the work is to be carried out and agree the surface levels of the ground and seabed. An agreed record of the levels shall be signed by the Contractor's agent and Engineer's representative and the Contractor shall supply the Engineer with three copies.

6.6. Pre-dredge Surveys (Clause 610).

A hydrographical survey over the are of the works was carried out in August /September 1991 by Jan de Nul NV in order to determine existing sea bed levels and to determine the presence of wrecks and obstacles in the areas to be dredged. The record drawing for the obstacle survey, drawing number SSS, is bound with under appendix 600/A, together with a schedule of interpretation. The survey by Jan de Nul, to be referred to as JDN survey lacks coverage in some locations of the sides of the channel.

6.7. Intermediate Surveys (Clause 614).

Intermediate surveys shall be at weekly intervals or such other intervals as the Engineer may require to monitor the progress of the work and ascertain the work done for valuation purposes. The extent of the surveys and the soundings line interval shall be that necessary to meet the requirements of the survey.

Intermediate surveys shall include areas (within or in the vicinity of the site) where shoaling is expected.

6.8. Post-dredge Survey (clause 615)

The post-dredge survey shall cover all the area dredged and shall extend to at least to at least 50m beyond the actual dredging limits. Check lines at right angles to the general direction shall be run at intervals not exceeding 200m. In areas where slopes are to be provided with temporary protection, the interval shall be reduced to 25m.

7. SURVEY DATA AND THE DREDGING DESIGN.

Survey data provided in the contract documents include:

- Shore control stations and their co-ordinates in Dar-Local system,
- Co-ordinates of channel centreline,
- Other data as provided on the drawings.

7.2. Corrections to the Channel Centreline Design

It was established while preparing to start dredging that the centreline as provided by the co-ordinates significant kinks, which were not intended by the designer were observed in the centreline. The survey consultants removed these errors and corrected the data to produce a centreline that was consistent with the contract drawings. The co-ordinates were corrected after redefining the centreline using a combination of straight lines and curves that best fit the centreline co-ordinates. The key centreline values are reproduced in Table 3 below.

Table 3 Key Parameters of Centreline

Pt	Co-ordinates		Chainage	Centre of Curve			Comment
	E	N	Kp	E	N	Radius	
IP 1	2294.421	1252.188	0.000				Start of CL
T P 2	2185.573	477.696	782.103				Start of 1stC
C P 3				799.198	672.538	1400.000	Centre of 1st C
T P 4	1680.247	-415.466	1832.792				End of 1st C
T P 5	466.666	-1398.205	3394.379				Start of 2nd C
C P 6				221.080	-1094.931	390.240	Centre of 2nd C

TP7	170.496	-148.879	3710.733				End of 2nd C Start of 3rd C
CP8				109.438	-1948.954	471.040	Centre of 3rd C
TP9	-305.508	-2171.877	4744.097				End of 3rd C
IP10	-157.347	-2447.650	5057.151				Intersection Pt
IP11	575.829	-4146.165	6907.151				End of CL

The width of the channel in all designated areas was determined from the centreline as defined by the above parameters by determining the position of the toe lines. The slopes were calculated to produce a slope of 1:5 except at the east and west ferry points.

7.3. Design of Dredging Work at the KOJ.

Surveys were also needed in making an accurate design of the Kurasini oil jetty (KOJ) main berth, and the coastal berth. Particularly, it was essential that the positions of the existing and proposed infrastructure be accurately determined. This involved a survey of the existing and berth and plotting the results on the detailed construction drawings of the coastal berth, which was then under construction.

8. CHART DATUM ISSUES IN THE CONTRACT.

A rule of thumb in surveying is that existing data to be used in subsequent surveys must be verified. The contractor was duty bound to carry out an in-survey and recomputed the in-situ volumes to be dredged to the satisfaction of the Engineer prior to the commencement of the works. He was therefore to carry out thorough datum checks.

8.1. The Confusion

There was confusion concerning the definition of the chart datum, caused by clauses 107 and 108 of the specifications. The requirement in the specification was to establish the relationship between admiralty and harbours *datum*, which confirmed chart datum to be 0.1 m above harbours datum. However, the British admiralty lowered their chart datum in 1992 from approximately MLWS o LAT, which in Dar Es Salaam differ by half a meter. The Engineer therefore ordered the contractor to use the pre-1992 datum as the datum for this contract since in the preparation of tender documents.

8.2. Consequences.

It is worth considering the ramifications of this decision as the contract drawings were mostly based on the JDN survey based on the pre-1992 datum. To lower the datum at this point in time of the project would mean increasing the dredge depth as well. This would increase the volume of material to be dredged and consequently the contract sum. As such the contract would be stalled pending an amendment and additional funds. The alternative would be to reduce the dredge depth by the same amount and consequently impose restrictions as to the type of ships to be expected in the port. This would, in turn mean that the economic benefits of the project would have to be re-appraised.

8.3. Datum for the Tide Recorder.

The dredging contract also required the contractor to install a specific design of tide recorder and tide boards prior to the start of any survey and to use the installed facility in subsequent surveys. The contractor installed 3 tide gauges, I automatic tide recorder and two radio tide gauges. One of the tide gauges had been installed at the police wharf alongside the automatic tide recorder. The recorder was set at a speed of 16 mm/hr. The second had been installed on beacon number 11 located in the vicinity of the middle shoal.

The datum problem was also encountered in setting the readings of the tide recorder that were to be used in conjunction with the soundings. Initially, the contractor had erroneously set the tide recorder to read harbours datum values. The in-survey was processing to bring the contours on the drawings to the per-1992 chart datum. All tide gauges were subsequently corrected to read chart datum upon completion of the in-survey and so remain throughout the project.

9. SURVEY OPERATIONS IN THE HARBOUR

Bathymetric charting is a process of acquiring and processing data on the underwater topography and its display in the form of chart. It is a type of surveying within the realm of hydrographical surveying and is unique in much the same way as dredging. Operationally the hydrographical surveyor has no contact, whether visual or otherwise, with the terrain being mapped. This particular feature determine, to a large extent, the methodologies used. Perhaps, the most spectacular ones are the absence of point marking and the uselessness of optical instruments that are a common feature in surveys on land.

The basic concepts of surveying, however remain common to land based and underwater surveys. Positioning in 3D is essential in the location of map features in both cases. Static surveys are replaced by dynamic surveys in water aboard a survey vessel. Positioning remains discrete in the time domain with sorter, almost continuous, intervals in the water. The position of an echo sounder mounted in the survey boat is determined first, with respect to selected chart, sounding and land based horizontal and vertical datum. The point is them transferred to the seabed by soundings that provide the depth-component.

Dynamic survey operations require an on bard computer for instantaneous processing of both the position fixes and the depth. Both the DGPS system and the echo sounder must be able to input their results to the on board computer. The software then combines the data and produces and on-line chart. Final and accurate processing is done in the office where hard copy charts can be printed and evaluation of the project performed.

Chart datum, being the datum plane to which charts are published is sometimes different from sounding datum. This is true for the port of Dar Es Salaam where the Harbours datum, marked by a benchmark on the wharf is often used for sounding. Work on the unification of the two must await a good knowledge of the local tidal phenomenon in the harbour. The tide tables currently in use are bur approximate as they are not predicted using the local tidal components of the Dar Es Salaam harbour. An analysis of the suitability of a chart datum depends on the analysis of water level of water level observations and subsequent tide predictions. Consequently, the reduction of the harbours datum to the admiralty chart datum must therefore be resisted as a general case but be made only in specific circumstances.

9.2. Position Fixing.

Position fixes of the echo sounder aboard the survey boat was made using the Differential global positioning system (DGPS) operation a set of real-time Trimble satellite receivers. At least one receiver was set up as a base station and operated continuously. Precautions for a continuous supply of power were made. The base station was equipped with a radio modem for transmitting data to survey boats and dredgers that are constantly in motion. The survey boats and dredgers were equipped with GPS satellite receivers- the rovers. Together, the static and roving stations operate in a real-time kinematic and differential mode.

On receiving the satellite signal, the static station broadcasts carrier phase GPS data to the roving receivers. The roving receivers move with the motion of the boat or ship receiving continuously the data relayed, through radio link, from the base station along with the signal transmitted directly by the GPS satellites in orbit. The algorithm in the roving receivers is designed to compare the two sets of data and determines the position of the boat or ship in real-time. The track of the survey boat was continuously displayed on the screen aboard the survey boat to enable the hydrographical surveyor navigate in strict conformity with the design.

Data processing involves the removal of ambiguities and ionospheric refraction errors whilst correction for multipath and tropospheric errors in the received satellite data. These correction for multipath and tropospheric errors in the received satellite data. These corrections enable the positions to be determined to within 30-50 cm accuracy. This accuracy is a marked improvement compared to non satellite systems employed in the past. It ensures that unwanted and unpaid dredging is reduced to a minimum as expected by both contractor and client.

The GPS shore reference stations, Life House(-360.393,-939.259) was set up on a point of known co-ordinates in a local system – the Dar local system. The point on was surveyed as an unknown point in a 5 – point GPS survey in which the other four were known. The resected point was verified by Engineer's survey team and approved for use. Other shore stations were surveyed either by resection or intersection employing a SET 3A Sokkia Total Station and levelled with respect to known benchmarks in the port area by the spirit levelling technique.

9.3. Sounding in the harbour.

Depth determination was designed in such a way that the survey boat was made to sail along lines of a given grid bearing as primary lines. A continuous sounding chart was produced on which tidal offsets were superimposed at known time intervals. Soundings for the production of bathymetric charts were recorded at 20m intervals. Also recorded simultaneously were position fixes and the corresponding tide. Secondary check lines were run at 40m intervals and at right angles to the primary lines. The contractor was to satisfy the condition that the depth obtained at the line crossings were equal. Bathymetric charts were produced and printed on engineer's demand at 1:2000 scale.

Sounding employed an echo sounder with a transponder mounted at the keel of the survey boat. Although equipped with dual frequency capability, the Engineer ordered that only the 210 kHz frequency was to be used in this project. The use of the 33 kHz frequency by the contractor could only be considered in rare circumstances. The echo sounder automatically registers the time interval between transmission and reception of a pulse of sound energy and its echo by a transducer. The time is transformed into distance and corrected for sea depth, beam width and salinity, among others.

Measured depths are further corrected for environmental effects such as sound velocity changes, waves air pressure and tidal changes. In depth determination by sounding it was important that the results of the measured tide be continuously relayed to the echo sounder aboard the survey boat. In is lowered on two strings marked at 5m intervals such that the measured depth is known. Calibration beginning and end of each workday by comparing sounded depths with ones measured using an echo sounder.

9.4. Determination of tide.

The admiralty tide tables show a consistent difference of approximately 0.5m with the THA tables, apparently due to the difference in datum. The possibility of local variations in tides also exists. A prudent determination of, not only the chart datum but also one for height systems on land can be arrived at by using the correct components when available.

Table 4. Tidal Components for Dar Es Salaam Port.

Name	Amplitude (A)	Phase (G)	Frequency (OM)	VO+U	F
Q1	0.025	39.5	13.398660883	104.1	0.814
O1	0.107	43.4	13.943035580	10.7	0.814
M1	0.010	14.8	14.496693943	89.4	1.519
K1	0.168	41.3	15.041068640	354.8	0.887
P1	0.063	43.1	14.958931360	37.6	1.000
3MS2	0.002	305.2	26.952312660	302.2	1.114
MNS2	0.007	122.2	27.423833743	66.5	1.075
MU2	0.020	106.4	27.968208440	333.7	1.037
N2	0.196	92.1	28.439729523	97.5	1.037
NU2	0.040	98.2	28.512583137	240.3	1.037
M2	1.078	113.3	28.984104220	4.1	1.037
L2	0.033	126.8	29.528478917	92.7	0.875
S2	0.537	152.2	30.000000000	35.0	1.000
K2	0.151	154.7	30.082137280	170.1	0.754
MSN2	0.002	71.0	30.544374697	301.6	1.075
2SM2	0.002	91.9	31.015895780	65.9	1.037
MO3	0.001	321.	42.927139	14.8	0.84

		7	800		4
M3	0.004	273.2	43.476156330	186.1	1.055
MK3	0.001	236.0	44.025172860	358.9	0.919
3MS4	0.002	267.3	56.952312660	337.2	1.114
MN4	0.005	284.7	57.423833743	101.5	1.075
M4	0.012	322.5	57.968208440	8.1	1.075
3MN4	0.002	331.4	58.512583137	274.7	1.155
MS4	0.004	13.2	58.984104220	39.1	1.037
2MSN4	0.002	59.4	59.528478917	305.7	1.114
3MK5	0.000	197.0	71.911244020	17.4	0.988
4MS6	0.001	60.8	85.936416880	341.4	1.155
2MN6	0.002	257.7	86.407937963	105.6	1.114
M6	0.004	285.9	86.952312660	12.2	1.114
MSN6	0.000	257.1	87.423833743	136.5	1.075
2MS6	0.003	342.5	87.968208440	43.1	1.075
2SM6	0.002	8.6	88.984104220	74.1	1.037
3MN8	0.001	119.7	115.392042183	109.6	1.155
M8	0.000	100.5	115.936416880	16.3	1.155
2MSN	0.000	124.	116.40793	140.	1.11

8		3	7963	6	4
3MS8	0.001	167.6	116.952312660	47.2	1.114
2MS8	0.001	156.8	117.968208440	78.1	1.075

The dredging project has made a great contribution in this regard by requiring that the contractor perform an analysis of the tides in Dar Es Salaam by computing the tide constants from tide recordings in the Harbour. Tides are primarily caused by the positions of the sun and moon in orbit relative to the Earth. Secondary factors include the shapes of the ocean basins and shore topography among others. Each harbour basin responds differently to tidal forces developing characteristically its own tidal wave. Under these circumstances oceanic tides at a point need be measured, analysed and predicted. The first two steps were undertaken within the dredging project. The results of the analysed measured tides are reported here in Table 4.

The contractor assembled tide recordings for the period ranging from 1st July to 10th December 1997. The computations from that data bay Delft Hydraulics using the GETIJSYS software were made before Christmas 1997 and provided 37 components for the tidal function. The results were compared with those from the semi-diurnal tides in the North Sea and gave satisfactory results in terms of standard deviation (0.609) of the residuals.

10. FUTURE TIDAL PREDICTIONS.

The dredging project has availed the surveying community with invaluable data and information. As more data is accumulated the possibility of own tidal predictions becomes real thus providing the ships that sail into the harbour with accurate depth information. Further, the possibility of getting a realistic datum for various height systems in Tanzania is now conceivable.

10.1 Discussion of the Computed Results.

The results given in Table 4 include all three groups of tidal components namely, the semi-diurnal, diurnal and long periodic waves. Only a few of the last group are featured understandably, because of the relatively short (6-month) record of data used. However, the first two groups are more important in tidal predictions at the sub-decimetre level.

The results show, as expected from theory, the dominance of the M2 followed by the S2 tides and further and to a lesser extent the N2, K1,K2 and O1 tides in that order. The M2 and S2 tides are semi-diurnal solar tides that would remain if the moon and sun moved in circular equatorial orbits. As also expected from theory their frequencies differ only slightly and the ration of their amplitudes M2:S2 is 2:1. The N2 semi-diurnal lunar tide is a result of the eccentricity of the orbit of the moon.

The K1, K2 and O1 are respectively luni-solar diurnal, luni-solar semi-diurnal and lunar diurnal tides that occur as a result of the inclination of the orbits of both the sun and the moon to the equator. Long periodic solar tides, especially the semi-annual tides are driven more by meteorological forces, seasonal variations in air pressure and solar heating than gravitational force. These will need a longer time series of data to be accurately computed. The non-linear terms such as the M4 occur in partly enclosed waters affected by the coastal configuration. These have also been computed

10.2. Significance of the Results

The computations presented here, constitute a major achievement in the history of tidal predictions and publication of navigation information that are relevant to the port of Dar Es Salaam. The tidal equation has as many as 400 known components. It is also known that of these, the 13 most important ones contribute 95% of the total tide potential at a point. Using the 37 components that have been accurately determined, the Dar port can now successfully predict tides with better accuracy than ever before.

One is reminded that the dredging project was meant to allow ships into the harbour round the clock all year round. Further, the *under-keel clearance* as given in the Dar Es Salaam *pilotage rules* in 0.7 m and will certainly require a very good knowledge of the Chart datum to be relevant. It will particularly, require accurate predictions of the tides to allow shipping in the channel at low tide. Predictions using global models for this purpose could lead to unpleasant situations since tidal components obtained using these models differ considerably from the actual tides as Table 5 below shows.

Table 5. Comparison of Tidal components in Dar Port obtained from measurements and those computed from global models.

Name	Amplitude Measured (m)	Predicted	Frequency Measured (°/hr)	Predicted	Type of Tide
O1	0.107	0.377	13.943	13.943	Diurnal lunar tide
M1	0.010	0.030	14.497	14.497	-ditto-
K1	0.168	0.530	15.041	15.041	Diurnal luni-solar
P1	0.063	0.176	14.959	14.959	Diurnal solar
N2	0.196	0.174	28.440	28.440	Semi-diurnal lunar
M2	1.078	0.908	28.984	28.984	- ditto-
L2	0.033	0.026	29.528	29.528	-ditto-
S2	0.527	0.423	30.000	30.000	Semi-diurnal solar
K2	0.151	0.115	30.082	30.082	Semi-diurnal luni-solar
M2	0.004	0.012	43.476	43.476	Ter-diurnal

The above table confirms the identity of tidal waves in frequency but different in amplitude. In other words, the arrival time of the waves is acceptable but the amplitude, in the global model will distort the expected height of tides in the port. To the port hydrographer, this means that the under-keel clearance envisaged while using the existing tide tables will not be accurate. These circumstances call for the Dar port management to use the computed values in all future predictions, considering also that the ships squat and possible additional draught due to its velocity could aggravate the situation.

REMARKS

11. ACKNOWLEDGEMENT.

Key participants in the project are: THA-Tanzania Harbours Authority as Client, HAM-Hollandsche Aannemming Maatschappij of Holland as Contractor and Evers Consult of Australia in association with Top-Carto of Dar Es Salaam as The Engineer. Project was accomplished under FIDIC-III terms and conditions of contract commencing on 26th May 1997 for an execution period of 39 weeks. Project was financed to the tune of USD 23.3 million and funded 60% by The Royal Netherlands Government (ORET programme) and 40%by THA. Material used in preparation of this paper in addition to personal involvement and experience include:

- The 1993 tender documents (vol. 1-5) for the Entrance channel improvements dredging works drawn by Scott Bertlin and updated in 196 by Evers Consult in association with Topo-Carto consultants.

- The 1994 report (vol.5) on the Dredging of the Dar Es Salaam submitted to THA by SLI Consultants of Canada.
- The 1996 tender evaluation report for the Dar Es Salaam port improvements dredging works by Evers Consult in association with Topo-Carto Consultants Limited.
- The 1997 Contract Agreement (vol.1 & 2) between THA and HAM for the Dar Es Salaam port entrance channel improvements – dredging works.
- Monthly progress (1997 – 1998) reports of the Engineer to the Client on the Dredging of the Dar Es Salaam port.

12. CONCLUSION

One of the reasons why the 1997/98 dredging project was accomplished on time and saved the client from performance related time extension demands by the contractor was <u>timely decision making</u>. Project management decision on contractor's performance on the under-water topography was based on hydrographical surveys and the analysis of bathymetric charts compiled from those surveys. The charts were produced by the contractor's survey department on instructions of the Project Manager and at the same speed as dredging itself. High speed was made possible firstly, by the highly qualified and experienced team of survey personnel in both the

Contractor's and Engineer's offices and secondly, by the automated state-of-the-art survey technology at disposal. Many engineering projects are denied these advantages and rely on ill-trained personnel and antiquated technology where a dividends for the client. It is here felt that survey services in engineering projects be entrusted to expert surveyors and, all projects that rely on survey services be managed in part by a team that include in surveys.

Often times visionary decisions can be made by consultants at design stage and incorporated into engineering projects with a primary purpose of providing checks or corrections, but with a secondary reason of <u>making rare information available</u> to the professional community also. Such was the case with the requirement for the Contractor to collect tide data as early in the project as possible and process part of it also as early as possible. The off shoots are that Tanzania mapping and charting community now know the tides of the Port of Dar Es Salaam from observations rather than theoretical models, and can predict tides that are relevant for the Dar Es Salaam Port and for the determination of heights country wide. Further the decision to provide THA with a set of high-tech hydrographical survey equipment for its own use is highly commendable. Such vision in consulting services ought to be encouraged especially in a third world environment crippled with shortage of funds.

The need to <u>cross-check and verify data</u> cannot be overemphasized. The Contractor had been informed in the specification to use admiralty chart datum, meaning the current one. The Engineer ordered the use of the previous (pre-1992) admiralty chart datum i.e. one in use at the time that the project was designed. The difference of 0.5 m would have been a point of contention between the parties. If ignored in favour of the specification the dredged material for the 6.9 km long, 140 m wide channel would have increased the payload by at least USD 2.5 million to the account of the client. It would have necessitated long extensions of time probably with some pay. It would have even required expensive arbitration. Similarly, financial and time losses would have been experienced on the project had the kinks caused by errors in the centreline co-ordinates been noticed during actual dredging.

THE END

PROJECT NO. 3: Singida – Shelui Road Section: Topographical Data Audit

FINAL REPORT, May, 2003

Foreword

In reading this report one must take the following two positions into consideration. Firstly, consideration shall be given to the fact that the terminology used here, is the same as that, which is used in the survey and mapping practice in Tanzania. It is possible therefore that a difference will be encountered which is at variance with the terminology used in the final report of the road design engineers, Techniplan. Care has however, been taken to explain the meanings in the text.

Secondly, one must also note that the focus of this assignment is on verification, or otherwise, of work already completed and whose report has been submitted to Client. Great effort has gone into providing adequate evidence, checks and balances to ensure that the findings and conclusions in this report are not misinterpreted.

EXECUTIVE SUMMARY

1. Objective and Scope of Work

The objective of the services rendered in this assignment was to check the topographical surveys carried out by M/s Techniplan along the Singida – Shelui road section, on which the detailed engineering design is based. In this regard, it was important to ascertain as to whether or not the coordinates and levels tie well with the actual positions at site and if they are tolerable for construction work.

The assignment was broken down into five steps to facilitate better management of the field-work. These are:

1. Site verification of all survey data including survey data published in Techniplan's final report to client.

2. Verify the planimetric positions of all centreline points, which include the vertices or points of intersection (PI), the beginning of curve (BC) and the end of curve (EC). Such verification is done with reference to points whose planimetric positions are known in a national system.

3. Verification of elevations of centreline points, namely, the PI, BC and EC at each curve. Such verification is done with reference to mean sea

level (MSL), aided by points in the vicinity of the project road for which elevations above MSL are known.

4. Replace centreline points, which have been established by Techniplan and, for some reason, have been displaced, mutilated, obliterated or broken to pieces.

5. In the cause of carrying out this assignment it was desirable to identify any missing information deemed necessary for authenticating the topographical data in Techniplan's final report to client.

In meeting the objectives of this study it was important and necessary to undertake the following operations:

1. Check on the horizontal (X,Y)-datum used by Techniplan in the computation of co-ordinates.

2. Check on the vertical datum used by Techniplan in the computation of elevations.

3. Physically locate points available in Singida – Shelui section, for the transfer of co-ordinates and elevations to the project area and also physically locate the survey points established by Techniplan along the centreline.

4. Replace any missing points along the centreline, which maybe found unusable.

5. Check the correctness, or otherwise of the determined control point positions using field measurements on a set of selected points.
6. Check on the correctness or otherwise of the elevations through determinations of elevation differences between selected points.

1.1 Available Data

The Singida and Iramba Districts are disadvantaged with regard to the availability of topographic data and information. A large portion of this area is not mapped and the other is depicted on old and outdated maps. Further the co-ordinate data in the area is reckoned to several datums that must be sorted out for each application.

1.2 Strategy for This Assignment.

The Consultant has opted to use the UTM Arc 1960 datum, the current national datum in Tanzania, for this assignment. A data search corresponding to the selected datum was undertaken within 20km of the termini of the centreline.

2. Singida Trig-Data

The area around Singida has trig-points established within 5-10 km from each other as a result of a series of triangulation survey of the area. The Consultant has identified for use only the UTM

Arc 1960 datum points and left out those which are reckoned to other datums such as TTM, Arc 1935, Arc 1950, New Arc 1950, etc. The coordinates and elevations of the twelve selected trig-points were searched from the master file for degree sheet 29 at the SMD. Part of this data was used to facilitate the consultant to arrive at a solution for this assignment.

2.1 Trig-Point Inspection.

The inspection of trig-points was carried out with a view of verifying their existence, physical state, accessibility and visibility with other trig-points on the list. The points finally selected for the assignment as a result of this inspection are Samamba Indaheya, Kikungula, Aerodrome Rock and Mantamuke (for elevations only). Aerodrome rock was partially used by Techniplan and is crucial to this assignment. Samamba is visible from it. Indaheya, though well placed, was found in error and subsequently removed from the list of usable trig-points. Mantamuke was used to confirm the perceived elevation error of Indaheya. Mantamuke was also adopted for height transfer checks to project road.

2.2 Inspection of Centreline Points

This inspection included the 57-control points established by Techniplan and the points defining the curves. This search was complicated by the fact that over 70% of control points are not plotted on the drawings supplied. In many places the car odometer was of great help but not along by-passes. The centreline points were identified and located or replaced, if missing. Because of vandalism of iron pins by villagers, wooden pegs were used most often for off-road points and iron pins or nails for points lying on the existing road.

2.3. Datum Checks.

Datum checks were made in order to verify the relative positions, both in plane and elevation, of selected points. The aim was to check on the work done by Techniplan. In locating these points as well as determining their elevations, three types of datum checks were made namely; between control points (BMs), between vertices (Vs), and between BMs and Vs. The results showed good agreement that lies within the margin of error of survey work in which Total Stations are used for distances and elevation difference measurements. These checks provide a vote of confidence on internal consistency in the data supplied by Techniplan.

<u>3.</u> Transfer of Elevations and Coordinates to Site

The exercise of transferring coordinates and elevations from trig-points to the site was aimed at determining similar values for at least one well located point, established by Techniplan.

3.1 Transfer of Heights

Transfers were done by establishing three tacheometric levelling lines passing through Techniplan's control point BM02. The lines are: Samamba – Aerock, Samamba - Mantamuke and Mantamuke – Aerock. As such, three elevations of BM02 were determined and the weighted mean of the three elevations was adopted to be the acceptable elevation of BM02, and used to compute all other elevations on this assignment.

The method used in the survey was one of reciprocal measurements of height differences between successive pass points using a total station. The elevations of Mantamuke and Aerodrome Rock were given the -1.67 m local correction before further computations were attempted.

3.2 Transfer of Co-ordinates

The objective of this transfer was to check on the positions of control points established by the design engineer and their validity. The transfer was undertaken through a short traverse that was oriented using the bearing from Aerodrome Rock to Samamba. The trig-points Aerodrome Rock and Samamba have good inter-visibility between them. Their coordinates published in the Arc 1960 Datum and on the UTM projection were selected to facilitate the co-ordinate transfer exercise. Aerodrome Rock was selected for instrument set-up because of its proximity to the site.

The transfer results show that all coordinates N(X) and E(Y) of the Consultant differ from those of the design engineer, indicating a shift in point positions determined by the design engineer from their would be UTM positions. The positions of all points established by Techniplan are therefore misplaced, if plotted on a map, due to an ill-definition of the (X,Y)- position of the initial point and orientation of the opening leg of the centreline (IP1-Vi) survey.

4.0   Findings,   Conclusions   and Recommendations

4.1. Major Findings

1. There is scarcity of mapping data and information in the project area. The three available Y742 map sheets (at 1:50,000 scale) cover some 45 of the 110 km corridor and are of a 1963 publication and hence grossly outdated. Another map available is the district

map of Singida and Iramba (uncontoured) of 1965 publication that was reprinted in 1968.

2. The official records for Trig-point data is kept in the file for degree sheet 29, which is the file containing Singida data at SMD. The records in this file show that local elevations in Singida are 5.4 ft (1.67m) higher than reality and must be corrected accordingly before use.

3. There is a trig-point located 2 km from the centreline in Singida from which coordinates can easily be transferred to the project road. This nearby trig-point is named Aerodrome Rock (Aerock) whose local triangulation coordinates have been transformed to UTM Arc 1960 datum. The use of elevation information associated with this point envisages that a correction of -1.67m will first be applied to its published elevation.

4. The trig-point Samamba, 102.X.1, reported as destroyed by the road design engineers and, by implication, unusable does exist. It constitutes a brass plate drilled in rock. This trig-point is visible from trig-point Aerodrome Rock.

5. The design engineer's drawings showing the topography along the corridor suffer from a number of weaknesses including;

- Lack of a scale bar. The scale bar is the only type of scale information for drawings that are likely to be reduced or enlarged through, say, photocopying as it shrinks and expands with such manipulations.

- Important data is partially plotted. It is seen that most of the 57-control points established by Techniplan are not plotted on the drawings. In the absence of adequate information on their location, these points cannot be located on the ground and are therefore not usable.

6. There is a significant difference between Techniplan's datum for the (X,Y)-survey data and the UTM Arc 1960 system now used in Tanzania.

7. The surveys carried out by Techniplan are internally consistent and therefore of expected standard but, are not accurate due to datum defects detected in all data. This data can be

made good thought transformation computations without the need for an all-out re-survey.

8. Topographical data showing position information agree well with one another. Datum checks on distances and elevation differences between BMs and PIs have confirmed this agreement, which shows the existence of internal consistency in topographical data acquired by Techniplan.

4.2 Major Conclusions

On Verification of the Horizontal Datum:

9. The horizontal datum used in the design of the Singida – Shelui road section of the Singida – Nzega road is a local datum to this project only. In this regard, the Consultant is unable to validate and verify the datum used in the road design against the national UTM projection for Arc 1960 datum. Users of these data particularly, planimetric data on all points of intersection, points defining curve parameters, and the initial and final points on the centreline, must take note of this fact when trying to integrate this data with data from other sources.

On Verification of Elevation Datum:

10. The consultant has established, from records kept and maintained at SMD, and confirmed by field measurements that the elevations of all local triangulation data in Singida are too high and must be reduced by a value of 1.67 meters to conform to elevations of the national triangulation data. The consultant has also established by measurements that the road designer did not take this correction into consideration in all design work. The elevation data of the design engineer cannot therefore be validated or verified. As a consequence, all the elevation data associated with the design of the Singida-Shelui road section of the Singida-Nzega road must be re-computed using the reduction factor of -1.67 meters to bring them into harmony with national elevation values. Points to be affected by this decision are: all centreline and cross-section points.

On Verification of Centreline Topographical Data:

11. The Consultant has upheld that there exists internal consistency in the centreline data. In other words, the location of one point relative to another in the project area is within acceptable margins of error. The need to relocate points established along the centreline does not arise at this or any other stage in the project. Established survey points must

therefore be left undisturbed. To this category of points (that must be left in position) are: the points of intersection, the points defining the curve parameters and the initial and final points of the centreline.

However, this seemingly good news has been blurred by defects in the horizontal and vertical datums and because of this fact the topographical data cannot be validated or verified.

4.3 Major Recommendations

12. It is the Consultant's established view that a disregard for the national control survey system and errors in the data held in Government custody is the major cause of problems encountered by the design engineer in this project. Other equally important problems include the quality of national survey data, its paucity and coverage. The roads sector ought to take note of this fact and implement the following recommendations for the future:

   - Set aside funds for the provision of geodetic (survey) control and its densification to a 3-5km level and to cater for topographical mapping of the route corridor as part of the cost of developing major road corridors.

- Appoint only experienced Tanzanian registered and licensed survey consultants, who are aware and appreciate the problems of surveys and mapping in Tanzania. A full list of such professional firms is obtainable by writing to the Secretary, NCPS, P.O. Box 9201, Dar es Salaam.

- Employ registered and licensed surveyors (survey engineers) among the rank and file of the engineers' establishment at TANROADS headquarters who will be charged with: the preparation of user specifications, liaisons with engaged consultants and contractors, advising TANROADS on all matters pertaining to surveys and mapping, supervising survey works, collecting survey data, keeping and maintaining records, etc.

Definitions, Acronyms and Abbreviations:

Arc 1935- a geodetic datum in Tanzania upon which N(X), E(Y) coordinates were computed and reckoned in 1935. The datum is tied to a triangulation chain that runs along the 30th Meridian, computed on Clarke 1880 (modified) ellipsoid.

Arc 1950 - a geodetic datum in Tanzania upon which N(X), E(Y) co-ordinates were computed and reckoned in 1950. The datum is tied to a triangulation chain that runs along the 30th Meridian and is computed on the Clarke 1880 (modified) ellipsoid

Arc 1960 - a geodetic datum in Tanzania upon which N(X), E(Y) co-ordinates were computed and reckoned in 1960. The datum is tied to a triangulation chain that runs along the 30th Meridian and computed on Clarke 1880 (modified) ellipsoid

BC — - beginning of curve

BM — - benchmark – the word is used by Techniplan to refer to a survey control point

DOS — - Directorate of Overseas Survey, now known as Ordinance Survey (OS)

EC — - End of Curve

FP — - Final Point on the centreline

GPS — - Global Positioning System

IP — - Initial Point on the centreline

MLHSD - Ministry of Lands and Human Settlement Development

MSL   - Mean Sea Level

PI   - Point of Intersection

RCSSMRS- Regional Centre for Services in Surveying, Mapping and Remote Sensing, now the Regional Centre for Mapping of Resources, RCMR

SMD   - Surveys and Mapping Division of MLHSD

TP   - Trig-point

TTM   - Territorial Transverse Mercator

UTM   - Universal Transverse Mercator

V   - Vertex (same as PI)

WGS84 - World Geodetic System of 1984 – parameters of this system were agreed upon internationally in 1984

Y742   - A world topographical map series published at a scale of 1: 50,000

1. THE PROJECT IN PERSPECTIVE

1.1.Introduction:

Topographical surveys of the land, whether performed using ground or aerial methods, are meant to provide a model of both landscape and relief to the end-user. In road works, such surveys enable the design and construction engineers to visualise and compare several alternative routes and subsequently, locate the selected one on the ground. Land based information and topographical data for the formulation of the envisaged model include data on terrain features and on land use. In this regard, it is the objective of survey works to determine the topography to appreciable detail, locate planimetric features and stake out the designed road on the ground.

The relief features depicted at the time of surveying the topography are crucially important in optimising on the resources including, the estimation and computation of the earthworks. It is most desirable therefore, that such surveys be performed with a high degree of responsibility.

Important tools in obtaining a realistic ground topographical model include the availability of adequate spatial terrain information upon which to base subsequent surveys. Spatial information is often available in the form of medium scale topographical maps, coordinates of survey markers, in well known reference systems, elevation of benchmarks and other survey control points, etc. It is expected that the maps to be used are current and the control survey points have been professionally determined and of adequate density per unit area within the project corridor.

1.2. General Quality of Data in Singida District:
A general study of the project corridor for the Singida-Shelui road section of the Singida – Nzega road shows that:

1. The area is disadvantaged with regard to the availability of topographical maps. The road corridor is supposed to be covered by 12 standard map sheets of the Y742 (or 1:50,000 scale) series, coded from 82/1 to 84/4. However, eight of these sheets in the Singida – Kiomboi area have never been published. An area of over 5,000 sq km falls in this group, and so is much of the Singida – Shelui road corridor. It is also noted that even the available map sheets are about forty years old. Some of the information on these sheets could, therefore, be misleading.

2. Trig.-point information (co-ordinates and elevations) in the project area is of a mixed bag. In this bag is the data published in the old territorial transverse Mercator (TTM) system. Some of the data is published in the universal transverse Mercator (UTM) system. Further, there exist errors in the data that require corrections. The corrections have been noted by the national surveys and mapping authority but have not been

effected. Great care must be exercised in the use of such data.

Notable among the available trig.-point information media is the existence of two trig.-point charts, which depict variable coverage and detail particularly in Singida and Iramba Districts. In particular, the more recent chart, published by the then Regional Centre for Services in Surveying Mapping and Remote Sensing (RCSSMRS) for East Central and Southern Africa based in Nairobi, has opted to leave out most of the trig.-points depicted on the 1961 chart published by the Surveys and Mapping Division, SMD. If the omission is deliberate then some trig-points on the 1961 chart could have been found to be of questionable quality and hence their omission from the more recent publication (c.f. appendices 7 and 8).

3. There are no levelling benchmarks in the Singida mapping block SB-36-4. Available elevation information is associated with existing trig.-points and was determined by trigonometrical levelling methods. The implication here is that the data upon which centreline elevations is to be determined are

only good to a decimetre, at most. In places where reciprocal determinations of heights was not done, then the effect of Earth's curvature and atmospheric refraction in that data, known to be of the order of 27cm per 2 km distance, is staggering. Unfortunately, a report on the determination of such elevations is not readily available to clear up such doubts. It has been presumed that reciprocal measurements were made in all cases, as the practice dictates.

1.3. **Expectations of the Design Engineer:**

Techniplan's report on Topography and Geodesy (annex 2 of the final report), is not detailed enough to be used for an analysis of the published data. However, from the appendices to the report one is able to note the following:

1. Techniplan, the design engineers of the Singida-Shelui road, expected to be provided with trig-point and benchmark data located within 5km to 10km of the road termini, when they were commissioned to undertake this work. Having failed to obtain such data, the alternative left was to improvise on whatever information was made available in the vicinity of the project. In this regard, some so called "official

benchmarks" were used even though some are not "accepted as geodetic stations" by the survey and mapping authority, SMD. Improvisation was undertaken to the extent of even using graphical derivations of horizontal coordinates upon which the road survey was to be based.

2. The design criterion adopted by Techniplan for accepting co-ordinate data was that such data was accurate to within a10 cm accuracy with that of RITES, the feasibility study consultant, and not that of the Government authority on surveys and mapping. The design engineer further ignores the official map projection used in Tanzania as "linkage to the UTM datum as envisaged by Techniplan's technical proposal of 1998 is not essential for the purposes of this project"

These two developments by the design engineers is consequential in that internal consistency of data is considered as overriding and it is not important as to how the surveys in this project relate to other surveys, and also how a plot of the surveys and designed road would appear on a map, if plotted using their data. It is noted here, that ignoring the UTM system leads directly to the introduction of possible shifts, rotations and scale errors in the co-ordinates of survey points.

3. The design engineer's discussion in appendix C of the final report to client, covering various data types on existing survey control and their source, are of little significance, if any, to the project. Practice dictates, and it is desirable, that a choice of a single survey control system upon which to reckon survey data be done before the surveys commence and such a system be used throughout the project and at all of its stages. Also desirable would be a possibility of transforming one system into the other, where a multiplicity of systems exists, so that only are of the systems is adopted and used. It is also of little significance to the project as to whether the data was obtained from the headquarters or regional offices of the Ministry of Lands and Human Settlement Development, MLHSD or any other source. A discrepancy in data is always a discrepancy and it is the cause of the discrepancy that is important in assessing data quality. Finally, and as a consequence, the conclusions at the end of the report indicate that the design engineer

opted to establish own local system or datum of surveys on this project.

1.4.The Way Forward:

A clear understanding of issues and problems always leads to a professional approach, which in turn enables a correct solution to be attained. The situation analysis made in sections 1.2 and 1.3 above has therefore been instrumental in understanding fully the root causes of anomalies perceived in the data on this project. In order to put the matter in proper perspective the consultant decided on the following general approach in addressing the terms of reference (TOR):

i. Made a thorough study of trig.-point data and information within the road corridor. The study was extended to survey trig.-point data in areas beyond this corridor, where necessary, whilst confining operation within the Singida and Iramba Districts.
ii. Used the official national datum in all work on the checking of surveys in this project. The official datum is the Arc 1960 datum. The datum is referred to the Clarke 1880 (modified) spheroid and the data published on the UTM map projection system.

iii. Applied independent checks on the data upon which the works were based particularly, the levels and on field measurements.

iv. Used simple methods of verification of data while avoiding the use of systems, such as GPS, which may not be merged with the Arc 1960 data without remnant shifts.

In applying this approach the consultant was able, at the outset, to identify possible weaknesses in the data and upon which initial investigations focused. These weaknesses emanate from the following sources:

a. Improper identification of both the horizontal and vertical datums (Arc 1935, Arc 1950, Arc 1960 local) for survey works on this project.

b. Confusion in the use of existing co-ordinate information, particularly the units in which the coordinates are given (metres, feet). It was appreciated that some of the data is over half a century old and has not been adjusted by the classic least-squares estimation method.

c. The projection used for the published data i.e. whether it is the Tanganyika Transverse Mercator, TTM or the Universal Transverse Mercator, UTM or any other.

d. The method used in the survey to establish survey control and elevation from Singida to Shelui namely, opening and closing a survey on points whose coordinates are determined by graphical scaling from another source whose credibility has not been established.

e. Dealing with adverse circumstances such as;

- Absence of topographical maps for an 80km stretch which is the larger potion of the defined route,
- Locating trig-beacons and other survey markers on the ground, in the absence of maps,
- Seemingly destroyed survey markers such as trig.-points.

2. INVESTIGATIONS ON SINGIDA TRIG.-POINT DATA

In this chapter the report focuses on the search and inspection of data from Government records and subsequent identification and location, on the ground, of trig.-points whose data had been found in the official files.

2.1 Search for Basic Trig.-Point Data

The trig.-charts discussed earlier in section 1.2 and the available maps of the area around Singida made it very promising for the consultant to start investigations. Appendices 7 and 8 show that Singida could have trig-points located within 5 km from each other. The consultant was keen to see the data pertaining to these particular trig.-charts, available maps and obtain any comments made by the custodians of the data.

Table 1: UTM coordinates in project area

POINT	N(X) In metres	E(Y) In metres	HT In ft/m	COMMENT
TP 222 Harara	9540 175.027	718 495.522	6,250.6	Primary
TP 223 Kianga no	9531 863.868	665 013.971	5,888.9	primary

TP 224 Mjagunda	9495 077.438	720 238.576	5,674.1	Primary
TP 225 Kikungula	9502322. 148	672 848.278	5,198.4	Primary
TP 226 Kingolani	9531 861.754	646 352.575	5,208.4	Primary
TP 227 Gila	9495 578 .993	645 424.784	4,914.0	Primary
102.x.1 Samamba	9450 580.4	694 593.5	5,525.3	Secondary
102.x.7 Indaheya	9468 740.673	675 210.044	5,085.4	Secondary
122.x.2	9435 636.9	720 195.9	5,684.3	Secondary
122.x.3	9468 020.523	691 200.214	5,639.4	Secondary
1012 Aerodrome Rock	9468 020. 523	691 200.214	1536.42 (m)	Transformed from TTM/local
1008 Mwanjoka	9464 029.024	703 801.086	1746.72 (m)	Transformed from TTM/local

A further search revealed that there were an additional two (2) trig-points whose data on the

TTM projection, were recently transformed to the Arc 1960 datum on UTM projection, bringing the total number to twelve (12). Table 1 above gives the N(X) and E(Y) Arc 1960 datum coordinates on the UTM projection of the twelve trig-points close to project road, around Singida and their elevations, as extracted from the master data file and including the two latest additions.

There is need to underscore, at this stage that, Table 1 is compiled on the basis of <u>availability</u> of UTM Arc 1960 data on file. It does not suggest, in anyway, that the points exist on the ground or give the state of the trig.-beacon at its place. It does not even suggest any level of accessibility or even closeness to or otherwise from the project road.

The file also provides other co-ordinate lists, as shown in appendices 10 – 12, of data available in the official file at SMD, for the Singida and Iramba District close to the area in question, known as "degree sheet 29" for which UTM Arc 1960 data exist. such as;

- Coordinate and elevations list of 14 other so-called secondary triangulation points. It is of note that only four of the trig-points in this list also appear in Table 1 (cf. appendix 11).

- Coordinates and elevations list of 32 points of the Bubu triangulation on the UTM projection. It is notable that this list includes only one point of Table 1 (cf. appendix 12).

Notable in these lists are two comments that have been taken very seriously. These are;

- *a comment inserted in red ink which states that "Singida local heights; add −5.4 feet for correct height (local figure too big)". This comment will be taken in its proper value while discussing elevation transfers to the project site.*

- *A second comment made which is noteworthy is that the coordinates of the Bubu and secondary triangulations are "for topo only", meaning the coordinates could be used in mapping but not other, more precise works.*

The search conducted at SMD and the observations made here led to some firm conclusions, namely;

1) That the coordinate data to be trusted in any non-mapping works in Singida are those supplied in Table 1, should they be identifiable on the ground. By the same token, no other

coordinate information is to be trusted. However, a physical inspection of these trig.-points, to establish the usability of each one was required as an essential step in this project.

2) The elevation data in Table 1 and those provided in other tables in the official file for degree sheet 29 may be used in all other works including topographical mapping once the −5.4 ft correction has been confirmed through measurements or otherwise.

2.2 Physical Inspection of Trig-Points

The search at SMD in Dar Es Salaam having concluded, the logical step to follow was to locate and inspect the trig-points on the ground and establish their usability, or otherwise, in this work. This exercise was aided by topographical map sheets 102/1, 102/3 and 102/4 and the Singida and Iramba Districts map published at a 1:250,000 scale. Towards the end, on the Shelui side, the road passes through the Wembere plains from Shelui village to the final point, F.P. In normal practice, trig-points in a rolling terrain are not sited on the plains. The nearest known trig-point in this area is TP 225 Kikungula, situated near Misigiri village at an approximate chainage of 74+800. Information obtained from the inspection of maps enabled the consultant to establish that only half of the points in Table 1 would be useful to the exercise of checking topographical data on the project site. These points and a brief description thereof are discussed in appendix 13.

The inspection report leads to these conclusions:

- There are six trig-points to be relied upon in the work of verifying Techniplan's topographical data on the Singida-Shelui road section of the Singida-Nzega road. In the terminus at Singida a key role in the work will play any two of the four trig-points identified. In using classical measurements a key factor, in selecting the trig-point to use, will play the inter-station visibility and distance.

- The notable difference of the elevation for 102.X.7, Indaheya, between the value (5085.4 ft) entered in the file and the value (5073 ft) shown on the map must be explained before either of the elevations was adopted for use. The alternative was to discard the use of elevation information from this trig-point. This difference of 12.4 feet or 3.78 metres is definitely a gross error that lies outside any margin of permissible error associated with survey measurements.

- Aerodrome Rock is a trig-point that is located closest to the terminus in Singida. It was also partially used by the design engineer in establishing the centreline. It is therefore the point that must first be verified in the transfer of elevations and be used thereafter if the verification is favourable. Visibility to a second trig-point was necessary if the co-ordinates and bearings of the Arc 1960 datum on UTM projection were to be transferred to the road centreline.

3. DATA SEARCH AND INSPECTION OF

TOPOGRAPHICAL POINTS

This chapter presents the results of the efforts made to locate and/or establish the survey markers erected by the design engineer on site, for various purposes, and verify data thereof. These survey points include; survey control (BM) points, Points of Intersection (PI), points marking other survey parameters such as beginning of curve (BC) and end of curve (EC), and points defining the centreline erected at 250 meter intervals on either side of the road and 50 meters away from the symmetrical centre. The data for these points is well documented in the report on "topography and geodesy" and "alignment design". The terms of reference for this assignment requires the consultant to " establish the road centreline as per existing design and setting out data". In other words, the data in the two appendices to the report are to be used unaltered.

3.1 Techniplan's Basic Survey Control Points.

The official report on "Topography and Geodesy" gives data and identification information for the 57 survey control points that have been established along the project site and determined to facilitate all other surveys on this project. It was therefore of utmost importance to locate these points, referred to by the report as benchmarks even though the terms of reference do not explicitly call upon the consultant to verify or otherwise data for these points.

The inspection results are discussed hereunder.

1.	The point named BM01 and used in locating the initial point, IP1 (defining chainage 0+00 of the road) was not located and identified on the ground. Its description of "nail in the ground" and the photo reference provided in the report was far from giving a clear identification of the location. However, the IP was later identified and established, as were the initial centre-line points located at 25m on either side of the IP1.

2.	The point named BM02 located "on the corner of the concrete", as per description and on the billboard showing direction to Saba Saba Primary School is easily identifiable and reachable. It served the field survey work well, as an easy connection to existing trig-points.

3.   Point BM03 is described as "a nail on the rock". Both the chainage and photograph led the team to identify the rock and particularly, the crack lines thereon. However there was no nail on this rock seen at the location described but a hole at which a white paint can be seen and is now fading. The field survey team adopted the drill in the rock (2-3cm in diameter) to be the point in the description.

4.   The point BM04 is most unfortunately placed, if at all it is where described as "right side after the corner'. The description does not point to any mark as is the survey practice, although an iron pin in concrete located nearby has been identified and adopted as BM04.

5.   The survey control point BM05 is not well identified on the stated culvert that was easily located by chainage. There is a rough cross mark however, that has been assumed to be the point in question as there is no other point,

along the centre of the culvert, as the photograph shows, that fits the description. The description ''right side, top culvert'' means right side of the road and on top of the culvert may go well for levels (culvert is flat topped) but not for position of the survey point.

6.   The description of BM06, like that of BM05, of "right side , top culvert" is not unique as required of a survey point. However, in this case, there exists a faint cross that was assumed to be the point in question.

7.   Survey points BM07 to BM12, as well as BM17 to the last of them, i.e. beyond the 34 km distance (Ch 34+00) are not indicated (mapped) on the drawing A.2.3 of the final report. It is therefore not possible to locate all these points without first taking the trouble of plotting the points accordingly. The points BM13 to BM16 exist on the design engineer's topographical map but not on site. Fragments of concrete on the side show that it was vandalised.

8. The design engineer's control point BM 21 was found to be in good condition and identified whilst BM 22 was doubtful. According to the description in the report, BM 22 was supposed to be on the side of a signboard. Upon inquiry to the local people it was found out that the original signboard had been removed and three new ones erected around the old one, creating confusion.

9. It was not possible to locate, with certainty, the final centreline point (FP) at chainage 108.149 in Shelui because all the BMs from chainage 74+800 onwards are not indicated on the longitudinal profiles and design engineer's topographical maps. There are so many marks at the possible location of the final centreline point, FP that one cannot, confidently, locate the proper end of the Singida – Shelui serve for the billboard marking the end of the road and start of the next. The area is waterlogged during the rains.

3.2. Centreline Points and Data Verification:

The initial point IP1 defining the road centreline was located and identified alongside the location and identification of initial points defining the centreline and located on either side of the road. The centreline point is an iron pin and the off-centre initial points are iron pins cast in concrete. This scenario remains unaltered throughout the road centreline. Having located the initial points, the exercise of locating the remaining points in these series continued well using the design data.

The approach taken in this exercise was to locate all points of intersection (PIs) in succession from chainage 0+000 to the end using the bearing and distance method of point fixation. Each point was fixed on both faces of the total station to desired accuracy. The terms of reference call upon the consultant to ''establish the road centreline as per existing design and setting out data''. The fixation of the curve points (BC, PI, EC) respond adequately to this call.

The centreline already established on the ground using design data will not require changing since the initial orientation of the centreline is independent of the coordinate system but adheres to the existing alignment of the first straight distance. Checks on this alignment were observed throughout the setting out exercise.

The following considerations were also taken:

- The points were located by setting out distances between vertices and angles of deflections. The curve points were also located by setting out corresponding parameters. Before the set-out data for the centreline could be used it was necessary to first compute the coordinates of all vertices using the extracted values. Results of this computation showed very good agreement in all aspects, proving that it was the right data to be used in the setting out exercise.

- The computation of "joins" (bearings and distances between successive points) has been done from coordinates of successive points of intersections (PIs) provided in the report particularly, the appendix on "alignment design". The computation of joins is a standard procedure and yields the bearing and length of each line defined by successive vertices. The difference between a reverse bearing to a vertex and that of a bearing to the subsequent vertex is a complimentary angle to the deflection (primitive) angle. The results of this computation confirmed that the line lengths are all equal to those provided in appendix A.

- An attempt has also been done to locate the main points on circular transition curves from survey control points (the BMs), established by Techniplan, starting with the vertices and including the off-centre points defining the centreline. In the latter case there was evidence of such marks having been uprooted and destroyed. It was established that the monuments were destroyed by villagers who wanted to make use of the iron pins cast in the cement. In many cases there remained only a hole with both the cement and iron pin carried elsewhere for ultimate destruction. It is highly recommended that an alternative method of monumenting survey points along the road alignment be used. As construction could be underway in the next 6 months or so, wooden pegs, which carry little commercial value, have been employed. Wooden pegs of 15-20cm lengths are good enough for the job at this stage in the design process. The work was accomplished side by side with the placement of curve points. At 50m perpendicular to line defined by vertices (IP) temporary pegs were placed. Two such pegs on either side of the centreline formed a line along which wooden pegs were placed using the on-line method at

250m. intervals and in succession from chainage 0+000.

3.3 Datum Checks and Location of Vertices:

Datum checks were made to establish the credibility of the data of one point relative to another. These checks are independent of the co-ordinate system used. Any professionally executed survey is assessed, in the end with respect to its accuracy. Once accepted, the accuracy level is then known to be above a certain threshold. This means therefore that datum checks should start with a small sample and proceed on to a larger one only if the data is not verified. This was the approach taken in this exercise. Datum checks can also vindicate points that have been disturbed. The latter is not an indicator of bad data. Sample points must therefore be selected carefully. This exercise started therefore by selecting and identifying points upon which datum checks were to be made. The results were entered in Table 2 and the following scenario is noted;

- The design engineer's control points BM11 and BM12 at the Iguguno bypass were located and identified. The Vertices; V57, V58, and V59, at the same by-pass were found missing. The missing vertices were then replaced after taking join computations using the coordinates of BM 11 with orientation to BM 12.

- A datum check on the design engineer's survey control points BM21 and BM22, by distance measurements was made. Measurements were made from BM22, which was found "in situ" (a cast pillar). The consultant's datum check measurements differed from the computed one by 1.30 metres. The 1.3 m difference, we believe, is due to misidentification of point of reference rather than an error in the design engineer's work.

- The distance from BM 22 to the vertex V57, which was found in place and in good condition, differed from the computed one by only 0.140 m. Further, the measured distance from V57 to V58 differed by 0.210 m from that computed from the design engineer's data.

Table 2: Results of Datum Checks

Station	Measured Distance	Computed Distance	Difference in (m)	Remarks
SINGIDA AREA				

BM 5				Reference is on top of culvert
	953.335	952.646	0.689	
BM 6				Reference is on top of culvert
IGUGUNO BY-PASS				
BM 11				found in good condition
	1581.897	1581.508	0.389	
BM 12				Concrete slab reference removed
MISIGIRI VILLAGE				
BM21				Replaced, Location approximated
	574.	573.73	1.274	Possibilit

	980	3		y of mislocati on of one point
BM 22				Found in good condition
	320. 391	320.25 1	0.140	
V57				Replaced point
	2140 .600	2140.3 90	0.210	
V58				Replaced point

These results are good, indicating that internal consistency exists in the elevation data produced by Techniplan.

TRANSFER OF HEIGHTS

This chapter seeks to establish, or otherwise, the existence of <u>external</u> consistency on topographical data. The consultants started this exercise by seeking to establish such consistency on the elevation information, bearing in mind the aspects of correction to local elevations in Singida and the differing elevations of Indaheya from independent sources.

The purpose of height transfer is to provide a datum for elevations on the project road now that internal consistency does not pose a problem. In this lies also the significance of the Indaheya-to-BM02 survey and subsequently, the Samamba-to-BM02 survey also. The transfer of heights from Indaheya has been undertaken to Techniplan's survey control point BM02. At the end, a loop of 8 points was closed so as to provide an internal check on the reciprocal tachometry survey accomplished. The loop closed within 8cm, well within the capability of the method, and instrumentation, proving also that the workmanship was good.

4.1. Establishing the Correct Elevation of Indaheya.

The use of the official elevation of Indaheya has posed some serious doubts and problems. Its height is provided in Table 1 as equal to 5085.4 feet but its authenticity is questionable for the following reasons;

- The height of the same point on the Y742 topographical maps (1:50,000 scale) is equal to 5073.

- Nowhere else does the point itself feature in the triangulation surveys of Singida District even under secondary triangulation where say, Samamba is also published.

- A nearby point, 102.Y.3, Mantamuke appears on the Y742 topographical map with height 5082ft. Indaheya is located about 6km away to the southeast of 102.Y.3 and at a zenith distance of 90.5 degrees below it showing that its elevation must be significantly lower than 5082 by a few meters (almost 10 feet).

- Tacheometric levelling from Mantamuke to Indaheya also showed that the latter lies 2.1 meters (6.9 feet) below that of Mantamuke. It is therefore obvious that the elevation on file is wrong and it could be a typing error, which in survey practice is known as a blunder. Under the circumstances the height of Indaheya as obtained from the degree sheet 29 file cannot be used in any survey work. The elevation printed on the map is the right one considering also the 5.4 ft correction for local elevations in Singida. The transfer from Mantamuke, when corrected for the local elevation correction, gives 5073.1 ft as the height for Indaheya above MSL and which compares very well with the 5073 ft printed on the Y742 map sheet of the area.

- A trig point elevation above the same datum obtained through different methods cannot have a random error (acceptable) of the magnitude given here. There must prevail a blunder in one of the two or both and no other way.

It is therefore tempting to accept the height of 5073ft Indaheya as the correct local elevation figure.

4.2 Corrected Local Elevations

The official file for degree sheet 29 has noted, that all heights published, in regard to local control points (excluding primary & secondary triangulation points in UTM), which include all spot heights shown on the Y742 topographical maps must be reduced in value by -5.4 ft. or -1.67 meters.

Three trig points of significance for this road survey project that must be reduced accordingly are therefore, Indaheya (at 5073 ft), Mantamuke (at 5079.0ft) and Aerodrome rock (at 5041 ft). Elevations for Aerodrome Rock and Mantamuke are as provided in the official file while Indaheya is as printed on the topographical map.

Table 3: Elevations and UTM Arc 1960 Coordinates in Singida.

POINT	N(X) (m)	E(Y) (m)	HEIGHT (m)	COMMENT
TP 224 Mjagunda	9495 077.438	720 238.576	1729.46	Primary
TP 225 Kikungula	9502322. 148	672 848.278	1584.47	Primary
102.x.1 Samamba	9450 580.4	694 593.5	1684.11	Secondary
102.x.7 Indaheya	9468 740.673	675 210.044	**1544.60**	Secondary
1012 Aerodrom e Rock	9468020. 523	691 200.214	**1534.78**	Transformed from TTM/local
1008 Mwanjoka	9464 029.024	703 801.086	**1745.05**	Transformed from TTM/local
Mantamu ke			**1546.43**	Elevation only used

.bold are corrected elevations

The corrected elevations therefore are:

<u>Indaheya:</u>
Published (on Y742) local elevation   5073.0 ft

Less correction   -5.4 ft

Corrected Elevation   5067.6 ft

Or
1544.602m.

Aerodrome Rock :
Published local elevation   5041.0 ft
Less correction   -5.4 ft
Corrected Elevation   5035.6ft
Or
1534.776m.

Mantamuke:
Published local elevation   5079.0 ft
Less local correction   -5.4 ft
Corrected elevation   5073.6 ft
Or   **1546.431m.**

The consultant inserted these corrected elevation values into Table 1 to arrive at a new table for all Works in the Singida area, considering also the proximity of trig-points to the terminus. The values to be used in this project are all entered in Table 3 above.

4.3 Transfer of Elevations Along the Samamba - Aerock Line

Tacheometric levelling of a 55 km network by reciprocal measurements to Aerodrome Rock, Mantamuke and Indaheya from Samamba provided very good closures. The closures of +12 cm to Aerodrome rock, -25 cm to Mantamuke and -52 cm to Indaheya as well as -37 cm on the Mantamuke – Aerock line have been critically examined prior to their application in this work. The examination showed that only three lines be used in subsequent work, namely; Samamba – Aerock and Samamba – Mantamuke and Mantamuke – Aerock, whose misclosures were distributed, as is the survey practice, to all determined height differences en route.

The result of distributing misclosures provided three values for the elevation of BM02, that lies on all three lines as 1500.67 m, 1500.92 m, and 1500.72 m respectively. Survey practice calls for a derivation of one elevation value as a weighted mean of all three elevations. The weighted mean, and which is <u>the accepted and therefore adopted elevation of the survey point BM02, equals to 1500.72 m.</u>

It is coincidental and worthy noting that so as to arrive at this figure from Techniplan's data, the elevation of BM02 must be reduced by 1.65 m (1502.37-1500.72 = 1.65m). <u>This correction is approximately equal to the -5.4 ft local correction stated in the official file for degree sheet 29 triangulation data.</u> It serves as a confirmation that the −5.4 ft correction is genuine.

The question to be considered here is the following. Couldn't the consultant have simply applied the -5.4 ft correction for local elevations obtained from the file for degree sheet 29 without the hustle of a 55km tacheometric levelling network? The answer lies in the basic principles of surveying practice where checks and redundant data must be used before data is accepted for a given application, considering particularly, the conclusions made earlier. It was doubtful, in this exercise, to immediately accept the −5.4 ft correction, as even SMD has not applied it to the data to which it refers. The direct application of the -5.4ft correction to local elevations without confirming information would have been done contrary to the practice and in contravention of basic survey principles. The 55 km levelling exercise has therefore confirmed the validity of the correction to be applied and validated other data for this work in addition to providing confidence to the consultant in recommending the application of the local correction to local elevation data in this project.

4.4 Corrections to Elevations

The datum used in Tanzania for elevations is the mean sea level, which is a level surface. A shift in the datum means and implies that the entire surface has been shifted linearly by -1.67 m parallel to itself. All elevations based on this datum such as the elevations of all survey control points and centreline points in the Singida – Shelui road project must therefore be shifted accordingly. Shifting the centreline datum by -1.67 m means therefore, that all elevations along the centreline must be reduced by this figure and all elevations for cross-sections must also be reduced accordingly. However, such a shift will not call for re-computation of the earthworks - a measure that would have been required only if the field measurements of elevation by the design engineer would have been found to be in error.

Is it necessary then, to examine the height datum again at the other terminus at Shelui? The answer is that such measures are not required since internal consistency in the double run tachometric survey of the design engineer has been established. The tacheometric method used by Techniplan employed this check and has been the approach taken by the consultant also who has checked on and established the existence of internal consistency through datum checks of selected sections of the surveyed road. The consultant has also carried elevation datum checks of selected sections from Singida to Shelui and compared these with Techniplan's survey data found "en route" (section 3.3).

4.5 Elevation Checks on Centreline Data

Checks have been undertaken on the elevation data published by Techniplan in its final report on "Topography and Geodesy" The standard method used was that of reciprocal tachometry using a total station in which each line segment (for which height difference is sought) is measured twice with the instrument and the reflector being interchanged in the reverse measurement. This method is renowned for almost completely removing the errors due to Earth's curvature and atmospheric refraction, that otherwise accumulates with increase in distances. The data in Table 4 below was obtained in the course of making the necessary checks on the centreline data. The differences in the fore and reverse measurements of the reciprocal elevations indicate the prevalence of refraction in the data, which is mitigated by the average in their absolute values.

Table 4 : Check Survey on the Determination of BM06 Elevation From BM02 by Reciprocal Determination

STA	DISTANCE	ELEVATION DIFFERENCE Dh			HEIGHT
		FORE	**REVERSE**	**CORRECTED**	
BM02					1502.37
	895.069	2.342	-2.43	2.386	
SH1					
	202.513	0.074	-0.098	0.086	
SH2					
	164.075	-0.962	0.912	-0.937	
BM03					1503.905
	811.54	-4.124	4.033	4.078	
BM04					1499.827
	1710.065	2.666	-2.937	2.802	
SH3					
	849.466	2.956	-3.080	3.018	
SH4					
	970.572	-2.710	2.440	-2.575	
SH5					
	774.22	-9.274	9.192	-9.233	
SH6					
	472.896	-3.820	3.766	-3.793	
BM05					1490.04

					5
	569.838	-4.026	3.94	-3.983	
SH7					
	387.455	-1.165	1.099	-1.132	
BM 06					148 4.93
	SUMS	-18.043	16.837	-17.440	

The agreement in the data is quite good. It can be concluded from Table 4 that the internal consistency of Techniplan's work in determining elevations is good and the need to perform more checks on the elevations does not therefore arise.

4. THE HORIZONTAL DATUM AND CO-ORDINATE TRANSFERS

This chapter addresses the chosen datum for planimetric survey data in this project. It also presents results of work done by the Consultant to compute data, of a selected sample of points, on the 1960 Arc Datum and UTM projection. The computation includes comparisons made with planimetric data presented in Techniplan's final report to client.

The existence of a multiplicity of datum's for the definition of the (N(X), E(Y)) coordinates along the route has been noted in section 1.2 and are summarised below. The datums are;

- The 1935 territorial system with coordinates published on the TTM projection in units of feet
- The 1950 Arc datum with coordinates published on the TTM projection in units of feet.
- The New1950 Arc datum with coordinates published on the UTM projection in units of feet or meters. This data was authorized for use vide survey technical circular no 4 of 1956 to facilitate tying cadastral survey work on the triangulation system.

- The 1960 Arc datum with coordinates published on the UTM projection system in units of meters.
- Local triangulation surveys of Singida whose results were published in the TTM system before the combined adjustments that led to the latest Arc datum adjustments.
- Transformed coordinates of points belonging to local triangulation schemes and transformed to the Arc 1960 datum on the UTM system using a similarity transformation that included other trig-points.
- Local Control Datum established by Rites / M–Consult Joint Venture in 1996 and partially used by Techniplan as discussed in their report on "topography and geodesy".

The 1960 Arc Datum on the UTM projection is the system adopted by the national survey and mapping authority for Tanzania. The TTM projection is an old one and is discouraged for use, by the same authority. All the other datums are local to Singida and Iramba Districts.

In view of the above, a decision was taken by the consultant to base all work in this project on the Arc 1960 datum using the coordinates as supplied by the division of surveys and mapping and published in the UTM projection system, in units of meters. To this data was included all other coordinate data, which have been transformed from other datum's to the Arc 1960 datum. Points in the project area which qualify for use in this contract are provided in Table 3.

5.1 Transfer of UTM Co-ordinates to Site:

The trig-points Aerodrome Rock and Samamba have a good inter-visibility. Their coordinates published in the Arc 1960 Datum and on the UTM projection were selected to facilitate this exercise. Aerodrome Rock was selected for instrument set-up because of its proximity to the site control points established by the design engineer and the validity of whose coordinates is, among others, the objective of the transfer exercise. The transfer was undertaken through a short traverse that was oriented using the bearing to Samamba from Aerock.

Table 5: Co-Ordinate Differences between Techniplan's and Consultant's Values

SURVEYOR	STATION	NORTHINGS (X) in meters	EASTINGS (Y) in meters
TECHNIP	IP1	9468064.	693156.

LAN CONSULTANT		321 9468111. 110	588 693155. 421
	DIFF	46.789	-1.167
TECHNIP LAN CONSULTANT	VI V1	9467938. 370 9467976. 366	692743. 928 692745. 615
	DIFF	37.996	1.687
TECHNIP LAN CONSULTANT	V2 V2	94681`55. 771 9468190. 511	692596. 933 692594. 000
	DIFF	34.740	-2.933
TECHNIP LAN CONSULTANT	V3 V3	9468370. 944 9468405. 516	692591. 198 692583. 632
	DIFF	34.572	-7.566
TECHNIP LAN CONSULTANT	V4 V4	9469307. 996 9469346. 320	692777. 883 692749. 827
	DIFF	38.324	-28.056
TECHNIP LAN CONSULTANT	V5 V5	94697678 .677 9469713. 878	692643. 972 692607. 449
	DIFF	35.201	-36.523

TECHNIPLAN CONSULTANT	BM02 BM02	9468386.890 9468420.641	692588.230 692579.341
	DIFF	33.751	-8.889
TECHNIPLAN CONSULTANT	BM03 BM03	9469617.720 9469654.343	692681.130 692645.494
	DIFF	36.623	-35.636

All care and necessary checks were made to ensure that the observations made were of acceptable standard. In this regard, angles were measured on two faces and two zeroes, with a mean square error of 2.5 seconds of arc. Distances were measured three times and meaned. In this way were eight traverse points fixed in plane including the initial point, IP1 of the centreline and vertices V1 and V2. The results of this traverse computation were also used to fix some of the design engineer's control points BM02 and BM03 as well as the vertices V3, V4, V5 and V6.

The fixation of Techniplan's point and centreline vertices made it possible to quantify the departure of Techniplan's data from their would be positions in the 1960 Arc Datum. The comparison speaks for itself in Table 5 above.

It is noted from this table that all coordinates N(X) and E(Y) differ, indicating a shift in point position determined by the design engineer from the UTM positions. This means therefore that the position of all point established by Techniplan would be displaced if plotted on a map due to an ill–definition of the (X,Y)-position of the initial point and orientation of the opening leg of the centreline (IP1-Vi). The coordinate transfer also enabled the consultant to show differences in orientations of the design engineer's survey lines work. Table 6 contains bearings of lines and their distances derived from the computations in both design engineer's and consultant's determinations.

Table 6:   Techniplan's Values Compared to Arc 1960 Datum Values

STATION	TECHNIPLAN	CONSULTANT	DIFFERENCES
	$\theta = xxx° yy'zz''$	$\theta = xxx° yy'zz''$	$d\theta = xxx° yy'zz''$
IP1			

	θ =253.01.37	251.47.57	dθ=1.13.40
	S=431.453	431.390m	dS=0.053 m
VI			
	θ =325.56.08	324.42.05	1.14.03
	S=262.432	262.384m	0.048m
V2			
	θ=358.28.24	357..14.21	1.14.03
	S=215.249	215.255m	-0.006m
V3			
	θ=11.16.02	10.01.05	1.14.57
	S=955.467	955.371m	0.096m
V4			
	θ =340.08.15	338.49.32	1.18.43
	S=394.128	394.170m	-0.042m
V5			

	θ =3.20.12	2.08.40	1.11.32
	S=746.588	746.612m	-0.024m
V6			
BM02			
	θ =4.18.59	3.04.10	1.14.49
	S=1234.331	1235.474m	-0.143m
BM03			

Most apparent in this table are differences in bearings of lines of about one degree and fifteen minutes, and a dismal difference in line lengths.

At subsequent stages of development in the Singida – Shelui road project centreline data that has been referred to the national grid (official national datum) will be required. The consultant has computed the UTM Arc 1960 co-ordinates of all vertices of the centreline, in addition to the initial and final points, and are appended to this report (appendix 14).

5.2 GPS Transfers:

The use of the GPS system has deliberately been avoided in this work in order to reduce misunderstanding in the analysis and interpretation of any discrepancies to the Techniplan's survey data. This decision was reached by the Consultant upon the discovery of usable trig-points close to the project area, contrary to what was reported by earlier reports.

It is known that the GPS system uses a different ellipsoid from the Clarke 1880 (modified) used in the Arc 1960 datum. The WGS84 system used by the GPS system can be transformed to the Arc 1960 datum and then to the UTM system using global transformation parameters. But, experience in Tanzania has shown that further shifts in coordinates are required after the 7-parameter transformation has been accomplished. Even after such transformation, it has not been possible to totally remove discrepancies between the two systems to facilitate error free checks on work that was done and presented in a system other than the WGS84 system. The GPS system would have been ideal for a fresh survey rather than a technique for check surveys requiring close tolerances as required in this contract.

5. FINDINGS, CONCLUSIONS AND RECOMMENDATIONS

This final chapter presents the findings, conclusions and recommendations of the consultant in this project. It is a synthesis of the work discussed in the preceding chapters.

5.1. MAJOR FINDINGS:

ON GEODETIC AND MAPPING DATA

1. The control data in the Singida area is sparse and published in different systems (geodetic datums). The systems must be brought into harmony by transformation prior to being used on the road project, if points within a single system cannot be utilized because of paucity or other reason.

2. There are no levelling benchmarks in the whole area constituting Singida and Iramba Districts. Elevations for use on this project can only be transferred from trig-points whose heights are, at best, good to a decimetre. Because of the rough accuracy in elevation data on trig-points, it makes little sense in using levelling as a method for supplying elevation data on the centreline corridor. The tacheometric method using a total station is as good as can be.

3. There is a scarcity of mapping data and information in the project area. The three available Y742 map sheets (at 1:50,000 scale) cover some 45 of the 110 km corridor and are of a 1963 publication, using 1959 photography with

coordination supplied on the UTM, new Arc 1950 datum and hence grossly outdated. Another map available is the small-scale district map of Singida and Iramba (uncontoured) of 1965 publication and reprinted in 1968.

4. The official records for Trig-point data is the file for degree sheet 29, i.e., the file containing Singida data at SMD, show that local elevations in Singida are 5.4 ft (1.67m) higher than reality and must be corrected accordingly before use. Furthermore, a secondary trig-point numbered 102.X.7, and named Indaheya, whose data is published in the degree sheet 29 data file has an erroneous elevation. The latter error is, by a balance of probability, a typographical one, which is pitiful as data in a master file are often assumed to be free from error.

5. There is a trig-point located close to the initial point in Singida from which coordinates can be transferred to the project road. This nearby trig-point is named Aerodrome Rock (Aerock) whose coordinates have recently been transformed to UTM Arc 1960 datum.

The use of elevation information associated with this point envisages that a correction of 1.67m will first be applied to its published elevation.

ON TECHNIPLAN'S GENERAL WORKMANSHIP

6. The trig-point Samamba 102.X.1, reported as destroyed by Techniplan and, by implication, unusable does actually exist. It constitutes a brass plate drilled in rock. It is located some 17km to the south of Singida town along the Singida – Manyoni road at Samamba village, on top of a high rock boulder. A target placed on this trig-point can be visible from trig-point Aerodrome Rock located within Singida town, close to project road. Aerodrome Rock is strategically located to facilitate transfer of topographical data to project area with minimum cost.

7. The placement of control points, known as BMs, along the project road was done in an haphazard way in that the longevity of the marks used was not always taken into consideration. The description is ambiguous. As a result of this situation

most of the 57 control points established by Techniplan cannot be unambiguously identified. The description of the basic control point locations is not accurate enough as to facilitate accurate identification of established points. Even some of the photographs provided for the purpose of point identification cannot, in the absence of other features, be helpful enough in resolving apparent fuzziness in such descriptions as "nail an the rock" or "right side top culvert".

8. The design engineer's drawings, A.2.2 and A.2.3 showing the topography along the corridor suffer from a number of weaknesses including;
    - Lack of a scale bar. For drawings, which are likely to be reduced or enlarged, the scale bar is the only type that shrinks and expands with such manipulations. The consultant noticed that the drawings in question were reduced by 50% from their original plotting. The scale of 1:2000 indicated on the map is actually false and ought to be 1:4000. The numbers on these drawings

indicate the contrary. Again as a result of this reduction words and figures have become too small prints to be legible.

- Important data is partially plotted. It is seen that most of the 57 control point, in particular BM07 to BM12, BM17 to BM21 and the rest of them beyond BM23, are not plotted. In the absence of adequate information on their location, these points cannot be located on the ground. These drawings need to be corrected for this shortfall.

9. Reports of the design engineer contained serious errors, which were not upheld in the field. A case in point is the engineer's implication of absence of trig-points in Singida. The Consultant's methodology had to change when the real situation on site was known.

ON TECHNIPLAN'S SURVEY DATA

10. Techniplan's (X,Y) survey data is based on a local datum that constitutes an approximation to, but not a transformation to, UTM Arc 1960 datum

system while the elevations are based on a local elevation that differs significantly from the actual elevations.

11. There is a significant difference between Techniplan's (X,Y) survey data and the UTM Arc 1960 system now used in Tanzania.

12. The surveys carried out by Techniplan are internally consistent and therefore of expected standard but, are not accurate due to datum defects detected in all data. This data can be made good thought transformation computations without the need for re-surveys.

13. Topographical data showing position information, where the points are found, agree well when connected to one another. Datum checks on distances and elevation differences between BMs and PIs have confirmed this agreement, which shows the existence of internal consistency in topographical data acquired by Techniplan.

5.2. MAJOR CONCLUSIONS:

ON VERIFICATION OF THE HORIZONTAL DATUM

14. The horizontal datum used in the design of the Singida – Shelui road section of the Singida – Nzega road is a local datum. This fact stands out even though the road designer tried and managed to bring the coordinate values of the initial point close to their UTM Arc 1960 values. The TOR requires the Consultant to ascertain to the client whether or not this data tie well with the actual positions at site. The only values that tie well with the actual positions are UTM Arc 1960 values. In this regard, the Consultant is unable to validate or verify the data used in the road design.

15. The actual positions of all survey points established by the design engineer depart from those used in the design while survey line lengths such as the distances between points of intersection, PI have been rotated/swung through a bearing of about $1° 15'$ from the would have been or actual direction on the ground. In this regard the consultant is unable to verify the design engineer's data.

ON VERIFICATION OF ELEVATION DATUM

16. The consultant has established, from records kept and maintained by the officer in charge of topographical and geodetic surveys in the Surveys and Mapping Division, SMD and has further confirmed by measurements that the elevations provided for local triangulation data in Singida are too high and must be reduced by a value of 1.67 meters to conform to elevations of the national triangulation data in the UTM Arc 1960 datum. The consultant has also established by measurements that the road designer did not take this correction into consideration in all design work. The elevation data of the design engineer cannot therefore, be validated or verified.

17. As a consequence of the findings in (4), all the elevation data associated with the design of the Singida-Shelui road section of the Singida-Nzega road must be recomputed using the reduction factor of -1.67 meters to bring them into harmony with national elevation values. Points to be affected by this decision are: all centreline and cross-section points.

ON THE VERIFICATION OF CENTRELINE TOPOGRAPHICAL DATA

18. The consultant has checked the topographical data published in the final report of the design engineer to establish its worthiness or otherwise in setting–out works and has upheld that there exists internal consistency in the data. In other words, the data of one point relative to another in the project area is within permissible error. However, this seemingly good news has been blurred by defects in the horizontal and vertical datums and because of this fact the topographical data cannot be validated or verified.

19. The design engineer has followed closely the existing road alignment. The orientation of the initial straight stretch of the centreline adheres to this criterion and is therefore independent of any co-ordinate system. Further, the setting out of the road centreline has been performed using design and not derived data and using local (relative) orientation. In this regard, the need to relocate points of the centreline in plane does not arise at any stage in the project. The consultant has

confirmed by field measurements that the setting out of road centreline parameters has so far been done well. Established survey points must therefore be left undisturbed. To this category of points that must be left in position are: the points of intersection, the points defining the curve parameters and the initial and final points of the centreline.

5.3. **MAJOR RECOMMENDATIONS**:

20. Following the findings and conclusions of this report, the client must set in motion measures aimed at rectifying the situation by considering remedies to the reported coordinate and elevation distortions in site works, in the short term and preventing similar problems in the long term.

21. It is the Consultant's view that a disregard for the national control survey system and errors in the data emanating from datum data and information is the major cause of problems encountered by Techniplan. Other equally important problems include the low quality of national survey data, its paucity and coverage. Such concerns have been

raised at various fora by the Institution of Surveyors of Tanzania (IST) and the National Council of Professional Surveyors (NCPS). A number of studies have also concluded likewise. These problems will remain with the survey data and map user community for the foreseeable future. The roads sector ought to take note of this fact and implement the following recommendations:

- Set aside funds for the provision of geodetic (survey) control and its densification to a 3-5km level, possibly as part of the feasibility studies, in developing major road corridors.

- Set aside funds for mapping of the road corridors as part of the feasibility study and road design stages, by undertaking recent aerial photography and mapping, specifically for this work. Technology now exists, even in Tanzania that can facilitate mapping within a couple of months.

- Appoint only experienced Tanzanian registered and licensed surveyors on survey and mapping tasks. Engagement of registered and licensed surveyors is a legal requirement under the Professional Surveyors Registration Act no. 2 of 1977. A full list of such professionals and their firms is obtainable by writing to the Secretary, National Council of Professional Surveyors (NCPS), P.O. Box 9201, Ardhi House, 4th floor, Dar Es Salaam.

- TANROADS ought to employ qualified and experienced surveyors (survey engineers) among the rank and file of its engineers. The appointees shall, among other assignments, prepare user specifications, supervise surveying contracts, advise employer on all matters related to surveying and mapping, maintain a database, liaise with engaged consultants and contractors, etc.

Useful Information on Trig.-points of the Singida – Shelui Road Corridor

102.X.1, Samamba

<u>General Information</u>: Trig point number 102.X.1 located on Samamba hill with elevation determined as 5525.3 ft above MSL Samamba hill lies some 500 meters from the Singida-Ikungi-Manyoni Road. Distance from Singida town is about 17 km to the South East. This point has parameters entered in official file at SMD. It is also shown on the 1:50 000 map sheet named Singida East and numbered 102/4. It is also shown on the 1:250 000 Singida and Iramba Districts Map.

<u>Search Result:</u> Samamba has been found and identified. Upon locating and inspecting this trig point it was found out that the instrument pillar for this point was missing having been destroyed long ago. Found in place however, is a brass plate that has been drilled in a solid rock boulder. In Geodetic practice, the brass plate is the mark to which the coordinates (N(X), E(Y), h) are reckoned. This trig point is therefore suitable for use as a survey control point without fear or worry. Observations from this point can be done quite easily by centring the

instrument on the brass plate. Observations to the point will require a visible mark to be placed and centred on the brass plate.

102.X.7, Indaheya

<u>General Information</u>:   Trig point 102.X.7, code named Indaheya is located at Maswaswa. Indaheya has an official height of 5085. 4 ft above MSL. It is located about 17km west of Singida town. The point is shown as a secondary trig-point on the 1:50000 map of Singida West identified also as sheet number 102/3, but the height on the topographical map differs significantly from that in the file . The point however, does not feature on the Singida and Iramba Districts map of 1:250 000, for unknown reasons. It appears in the official file with UTM co-ordinates published.

<u>Search Result:</u> The point has been located and identified at Indaheya. This point was found in place and in good structural condition, though needed painting for good visibility and identification. The survey team has painted the pillar in white colour for identification and to facilitate easy observation from distant points. It is, for all practical purposes, in good condition for use in coordination and transfer of heights.

TP 225 Kikungula

<u>General Information:</u>   Trig point TP 225 code named KIKUNGULA is a Primary Trig point mapped at the north western fringes of the 1:50 000 map sheet 102/3 of SEPUKE. Its official height is 5198.4 ft above MSL but is mapped at 5198 ft on existing topographical maps. It also features on the 1:250,000map of Singida and Iramba Districts. The official UTM Coordinates and height are:

<u>Search Result:</u> The results of the first search for this point were not successful as the northern part and surrounding area to its north is not mapped on the Y742 maps.   However, concerted effort, using the local population was fruitful for the point to be located. It is in good usable condition.

1012 Aerodrome Rock

<u>General Information</u>: This is a secondary trig-point of the Singida local triangulation network whose data is given in the degree sheet 29 file on the TTM projection. This data was transformed recently to the UTM Arc 1960 datum in a similarity transformation computation that involved a resurvey of TP 224, Mjagunda and TP 225 Kikungulu. It is placed on a top of a rock boulder some 300 m., close to the Singida aerodrome and hence the name.

<u>Search Result</u>: The search was guided by an existing topographical map of the area at a scale of 1: 50,000 and local information, as the aerodrome is shown on the map but the trig-point is not. The rock is however, easily identifiable. The found trig-point constitutes a brass plate that has been drilled in a solid rock boulder. It is without an observation pillar and the instrument or target must be erected and centred on the mark on top of the rock during measurements. Aerodrome Rock (Aerock) is intervisible with trig-point 102.X.1, Samamba. It is ideal in the transfer of UTM Arc 1960 co-ordinates to the project site, due to its

accessibility and proximity to the site, at less than 1.5 km away.

1008 Mwanjoka

<u>General Information</u>: Trig-point 1008, Mwanjoka is a secondary point of the Singida local triangulation. Its data is obtainable from the official file for degree sheet 29 being published on the TTM projection. However the point now has known UTM Arc 1960 data after the similarity transformation computation that produced similar data for the trig-point Aerock.

<u>Search Result:</u> There was no benefit to be gained in searching for this point after identifying the trig-point Aerock and particularly after establishing visibility between Aerock and Samamba.

COMMENTS ON DRAFT FINAL REPORT FOR CHECKING OF TOPOGRAPHICAL SURVEY DATA FOR SINGIDA – SHELUI ROAD.

General comments

Generally the quality of report submitted by the consultant is good. However, the consultant has to give his opinion for their recommendations of corrections of elevations by –1.67 m.

<u>RESPONSE:</u>

The Consultant acknowledges the Client's satisfaction with the report. Further, in survey practice any figure that has been confirmed by independent checks is regarded to have been confirmed <u>beyond the shadow of the doubt</u>. The Consultant's opinion that the –1.67 m correction to all local elevations be respected by all stakeholders on this project is firm and unequivocal. Confirmation was done in section 4.3 of the report as also quoted here: "It is coincidental, and worthy noting, that to arrive at this figure from Techniplan's data the elevation of BM02 must be reduced by 1.65 m (1502.37-1500.72 = 1.65m). <u>This correction is approximately equal to the -5.4 ft local correction stated in the official file for degree sheet 29 triangulation data.</u> It serves as a confirmation that the –5.4 ft correction is genuine.

The question to be considered here is the following. Couldn't the consultant have simply applied the -5.4 ft correction for local elevations obtained from the file for degree sheet 29 without the hustle of a 55km tacheometric levelling network? The answer lies in the basic principles of surveying practice where checks and redundant data must be used before data is accepted for a given application, considering particularly, the conclusions made earlier. It was also not prudent, in this exercise, to immediately accept the −5.4 ft correction, since SMD has not applied it to the data to which it refers. The direct application of the -5.4ft correction to local elevations without confirming information would therefore have been done contrary to the practice and in contravention of basic survey principles. The 55 km levelling exercise has therefore confirmed the validity of the correction to be applied and validated other data for this work in addition to providing confidence to the consultant in recommending the application of the local correction to local elevation data in this project."

Detailed comments

1.   **Corrections to elevations.**

Under chapter 4 of the draft report, it has been reported that the local correction stated in the official file for degree sheet 29 triangulation data is −5.4 feet, which is about −1.67 m. Techniplan did not apply this correction from BM02 and thus it is true that all elevations are to be reduced by −1.67 m. Under para

6.2 major conclusions, conclusion no. 17 states that points to be affected by the above reduction are centreline and cross-section points. Furthermore, the conclusion states that it is advisable that the entire road profile be re-examined for a re-computation with regard to the earthworks. The consultant is required to confirm/state whether the reduction of −1.67 m has effect or no effect to earthworks.

<u>RESPONSE:</u>

The issue is covered under section 4.4 as quoted here: "Shifting the centreline datum by -1.67 m means therefore, that all elevations along the centreline must be reduced by this figure and all elevations for cross-sections must also be reduced accordingly. However, such a shift will not call for re-computation of the earthworks - a measure that would have been required only if the field measurements of elevation by the design engineer would have been in error."

2.   Submission of survey data, logbooks and original of maps and other survey data in electronic and hard copy.

Under Paragraph 4.4 of pre-contract discussions between TANROADS and the Consultant, it was agreed that, field survey data, logbooks and original of maps and other survey data will be provided by the consultant in electronic and hard copy. The consultant will be required to submit to TANROADS the field survey data, logbooks and original of maps and other survey data in diskette and hard copy during submission of the final report.

RESPONSE:

This contractual obligation will be obliged to in the final report to Client. Some of the data will form part of the appendix to the report. There are no diagrams/maps drawn by consultant for submitting to Client.

PROJECT NO. 4: CADASTRAL SURVEYING OF FORMER TPA HOUSING ESTATES AT DAR ES SALAAM, TANGA & MTWARA

'DRAFT' AGREEMENT

THIS AGREEMNT made on the …. day of …………… 2005 between TOPO-carto Consultants Limited of Post Office Box Number 35169 Dar Es Salaam (hereinafter called ''the Consultants'') of the one part and TANZANIA HARBOURS AUTHORITY of Post Office Box 9184, Dar es Salaam (hereinafter called ''the Client'') of the other part.

WHEREAS the houses forming the Client's housing estates at Dar es Salaam, Tanga and Mtwara have been transferred to the Government, specifically Tanzania Building Agency for the purposes of selling the same to T.H.A employees. AND WHEREAS the client has agreed to undertake on behalf of the Tanzania Building Agency, cadastral survey of the housing estates attached hereto as Appendix I so as expedite the process of procurement of Certificates of Titles by the new owners.

AND WHEREAS the Client is desirous and the Consultant is willing to certain consultancy services, viz, to conduct cadastral survey for certain plots belonging to the Client as stated herein.

NOW THEREFORE IT IS AGREED as follows:

1.0 The Client hereby appoints and the Consultant accepts the appointment for the provision of Consultancy services hereinafter specified subject to and in accordance with terms and conditions hereinafter set forth.

2.0 The Consultant shall, in accordance with the survey instructions to be issued by the Office of the Commissioner for Lands, Dar es Salaam and to the satisfaction of the Client and Director of Surveys and Mapping, perform THE CADASTRAL SURVEY OF FORMER TANZANIA HARBOURS AUTHORITY PLOTS as detailed on the schedule herein attached as APPENDIX I (hereinafter called "the Consultancy Services") which shall be read and construed as part of this Agreement.

3.0 In consideration of the services to be performed the client agrees to pay the Consultant the total contract price of Tshs xxxx (hereinafter called the "Contract Price ") in the manner and subject to conditions hereinafter set out.

4.0 That the total contract price specified under Clause 3.0 herein above shall be paid to the Cosultant in the following instalments subject to the conditions specified herein

PAYMENT SCHEDULE

a). 20% of the Contract price equal to xxxxxx Tshs Shall be paid to the Consultant by the Client, upon submission to the Client of Invoice and Inception Report covering the three sites of Tanga, Dar Es Salaam and Mtwara within three Four weeks of commencement of Works.

b). 60% of the contract price equal to Tshs xxxxxx shall be paid to the Consultant by the Client, upon submission to the Client of Invoice and copies of the unapproved survey Plans submitted to the Director of Surveying and Mapping from the Ministry of Lands, and Human Settlements.

c). That the remaining 20% shall be paid upon submission to the Client of Invoice and a copy of the survey plans approved by the Director of Surveys and Mapping in the Ministry of Lands, and Human Settlements Development.

PROVIDED THAT the total contract price herein before specified shall not be adjusted or varied upwards for any reason whatsoever.

5.0 The Consultant shall commence the consultancy services on the date to be mutually agreed upon and as notified in writing by the Client.

6.0 The Consultant shall proceed expeditiously and with due care in the carrying out of the consultancy services and shall complete the same within six months from the date of notification aforesaid (herein called "the completion period")

AND PROVIDED THAT

a). The aforesaid completion period may only be adjusted, upon proof that rainy days have impeded the progress.

b). Should the Consultant exceed the agreed completion period, the Client shall not be required to pay any extra costs. Any such extra costs shall be borne by the Consultant.

7.0 In the event the consultancy services are not performed in accordance with the said survey instructions and/or the said survey is found unsatisfactory by the Client and/or the Director of Surveys and Mapping on any other account, the Consultant shall re-perform the same at its own cost and expenses.

8.0   The Consultant shall take out maintain an insurance policy for its employees to cover them against any risks of damage, loss, injury or death while such employees are entering in, on and about or leaving the premises of the Client. The Client shall be fully indemnified for the claims, demands, actions, cost and expenses arising from and or connected with such damage, loss, injury or death aforesaid.

9.0   The Client shall be entitled to terminate this Agreement upon issuing of one-week notice, in the event that the Consultant fails to commence work for a period longer than one week, from the date of notification as aforesaid or should the Consultant fail to progress and complete the consultancy services for a period longer than two weeks after the due date of completion.

AND in the event that the Agreement is so terminated, the Consultant
shall reimburse to the Client any sum of money already paid including any
extra costs that the Client may incur in recruiting alternative consultancy
services.

10.0   In the event of a dispute or difference touching on the construction of this Agreement or the rights and obligations of the parties herein, unless the same is settled amicably, either party may refer such dispute or difference to two arbitrators to be appointment by the Chairman of the National Council of Professional Surveyors for adjudication and settlement in accordance with the Tanzania Arbitration Law.

11.0   Any notice or communications as required or authorized to be under this Agreement shall be sufficiently given or made if it sent by registered post to the ordinary address of the party or delivered by dispatch and signed for its receipt.

a).   In the case of the Client, addressed to the Director General for the attention of the Director of Engineering and Technical Services.

b).   In the case of the Consultant, addressed to the Managing Director, Topo-Carto Consultants Ltd.

12.0   This Contract is made in Tanzania and shall be subject to and construed in accordance with the Tanzania Law.

IN WITNESS WHEREOF the parties hereto acting through their representatives duly authorized thereunto have executed this **AGREEMENT** on the day and year first above written.

SIGNED and **DELIVERED** for and behalf of

TANZANIA HARBOURS AUTHORITY
NAME ....
SIGNATURE ....
DESIGNATION ....

In the presence of
NAME ....
SIGNATURE ....
DESIGNATION ....

SIGNED AND **DELIVERED BY:**

SIGNED: ....
NAME: Dr. Furaha N. Lugoe
QUALIFICATION: Managing Director

In the presence of:

SIGNED: ....
NAME: Miss. Adelina Kingu
QUALIFICATION: Personal Assistant to MD

APPENDIX I

S/n	STATION	DESCRIPTION OF ESTATE TO BE SURVEYED
1		Msasani Estate Type IA
2		Msasani Housing Type II & III
3		Msasani Plot No. 1826
4		Plot on Kimweri Rd.
5		Plots 577 & 569, Mindu St.
6		Plot 269 Alykhan Rd.
7		Plot 2114 Blk'A' Sea-View
8		Plot 2115 Blk 'B' Sea View
9		Plot 2115 Blk 'C' Sea-View
10		Plot 2104 Ocean Road
11		Plot 454 & 446 Charambe St.
12		Plot 95 Guinea Rd
13		Plot 194, Upanga
14		Plot 272 Upanga
15	DAR ES SALAAM	Chang'ombe
16		Plot 723 Upanga
17		Kigamboni Estate
18		Kurasini Housing Estate
19		Mgulani Housing Estate
20		Msimbazi Housing Estate
21		Plot 43/33 Kilwa Rd.
22		Gerezani
23		New Usagara Estate
24	TANGA	Chumbageni
25		Nguvumali/Kisosora
26	MTWARA	Ligula Plots
27		Railways Estate

Tanzania Harbours Authority

Office of the Director General

One Bandari Road Kurasini   P.O. Box 9184
Dar es Salaam

Telephone: (255-22) 2110401 Fax (255-22) 2113432

E-mail: cer@tanzaniaports.com

Please quote: EN/1/3/1

The Secretary & Legal Officer,
DAR ES SALAAM

CADASTRAL SURVEY OF FORMER THA HOUSING ESTATES

As you may be aware some of the THA houses sold to staff have no separate certificates of Title as they are part of estates whose ownership is evidenced by a single document.   In order to facilitate individual ownership, the Estates have to be subdivided and resurveyed to enable deed plans be prepared for inclusion in Certificates of Title.

In order to hasten the titling process THA has agreed to assist T.B.A. in carrying out the survey.

We have therefore evolved a draft tender document for the work which is forwarded herewith for your vetting.

Eng. F.A. Maro
DIRECTOR OF ENGINEERING
AND TECHNICAL SERVICES

NECESSARY DOCUMENTATION TO BE CONSIDERED AT INCEPTION

Survey Instructions:

The Project Agreement documentation calls for the Consultant to deliver the services in accordance with Survey Instructions of the Commissioner for Lands. Licensed surveyors are required by law to additionally obtain survey instructions of the Director of surveys and mapping. The Consultant has accordingly applied for the instructions from both the Commissioner and the Director of the Surveys and Mapping Division. It is worthy of note that the survey instructions of the Commissioner have traditionally been preceded by an application for permission to divide the land. At the same time the survey instructions of the Director of SMD await those of the Commissioner for Lands. Where problems are not anticipated surveyors have often proceeded with surveys pending the issuance of survey instructions. The Client is therefore requested to offer assurance that the project will not be hindered by a denial of survey instructions for this project emanating from a reluctance or otherwise of the CoL to grant permission to divide the various estates which are under title.

Permission to Subdivide Land Under Title:

This is a provision of section 83 of the Land Registration Ordinance (Cap 334) that has provided official form L. R. 30, being an application for the "Division of Registered Land".

> "The Registrar may, on the application in a prescribed form made with the consent of the Commissioner for Lands, of the owner of Government Leases or of a Right of Occupancy over any parcel, divide such parcel into two or more parcels by canceling the folio or folios of the land register relating thereto and preparing new folios in respect of the new parcels:"

Such an application must contain an attachment, being a plan of the subdivision. The Consultant has prepared such plan for each of the estates based on large scale (1:2,500) township maps and recent high resolution ortho-photo maps of the Dar Es Salaam estates and environs.

In using the large scale topographical maps, the maps were first enlarged to provide a clearer picture of the arrangement of the buildings and infrastructure as made by the engineers and as existing at the time of preparing the maps. Any recent developments were captured on orthophoto maps. In some instances, lay out plans of the engineer/developer exist and these have also been used in the preparation of subdivision plans and will be used in the staking out of plots during cadastral survey field work.

There are however some estates for which engineer lay out plans either original or 'as made layouts' have not been handed over to the Consultant. The Client is requested to expedite their delivery to the Consultant.

In addition to the use of these documents (layouts, maps and plans) in guiding the determination and survey of plots, they are also made to assist in the application for permission to subdivide the estates. But, as the Land Registration Ordinance (section 83) states, it is an application of the 'owner of the right of occupancy' and ought to be made therefore by the Client. The plans are provided herewith as appendix

No. 1 to assist the client in lodging applications for permit to subdivide the various estates. Also provided in this documentation are the relevant orthophoto maps.

Should the Commissioner accept the application and grant permission to subdivide the estates, as it is expected, the original titles and corresponding registered survey plans will be cancelled paving way for new survey plans to be registered and new certificates of occupancy to be issued.

Land Parcels of Semi-detached Houses and Adjoined Apartments:

Many houses listed for cadastral surveying in this project constitute semi-detached dwellings in which a wall is shared between two adjacent neighbors. The Town Planning Section of the Human Settlement Development Division (HSDD) has always insisted that a land parcel (or plot of land) must be an area of land under homogeneous property rights and unique ownership. Accordingly, where a wall is shared, the land parcel ceases to provide that homogeneity and uniqueness. In the absence of a Condominium or Strata Titles Act in Tanzania, the Town Planning authorities do not recognize a shared wall to be part of a plot boundary. Consequently, semi-detached houses and adjoined apartments can only be occupied in common as per Part XII of the Land Act No. 4 of 1999.

The Town Planning description leans more towards building regulations and building permits in urban areas and not on boundary delineation and demarcation. On the other hand, and from the cadastral survey point of view, land parcel or plot boundaries are defined by thin lines between centers of corner beacons or other markers and, can be defined irrespective of the envisaged property development or mode of occupancy of adjacent land parcels. However, the purpose of conducting a cadastral survey in a subdivision is to eventually obtain a land title for the land parcel in question and it is particularly for this reason that the position of the HSDD be seriously considered. The Client is assured of cadastral surveying of plots for individual buildings but not the partitioning thereof unless a consensus is arrived at to the contrary.

It is the exclusive advice of the Consultant that in view of the likelihood of a Sectional Titles or Condominiums Act to be enacted in the foreseeable future, itself being provided for under section 6.1 of the 1995 Land Policy document, a representation be made to the Acting Director for HSDD and the Commissioner for Lands and a permission be sought for the two departments to accept a shared wall as part of a boundary between developed plots or land parcels. Such acceptance by the HSDD can pave way for an inclusion of a condition in the Certificates of Occupancy of the neighbors sharing a wall for its safeguard as a recognized boundary between their plots.

The above advice must be given serious consideration if individualized titles are to be obtained for all those units in shared buildings. It is further noted that the majority of houses and blocks of flats belonging to the Tanzania Buildings Agency and probably the National Housing Corporation will benefit immensely from such a provision. In the absence of an agreement on this issue subdivision will follow the current practice and the current legal provisions. The plots will be surveyed for each buildings and titles in such buildings will be titles in common.

Awareness on Property and Land Rights
The reconnaissance tour of the Consultant in the company of the Client's Chief Estates and Rating Surveyor, Mr. J. M. O. Safari, has revealed that many occupants of the houses earmarked for cadastral surveying are busy with extensions, fencing, wall building and landscaping, among other activities to the land around their houses. In some instances, these developments have encroached on areas set aside for utilities while in other cases the side roads have been

blocked altogether. In some instances, the neighbors are unaware that they are likely to be tied up in a common title with neighbours. Many of the small houses have shared pit latrines located between houses, etc. It is prudent that they be informed of the real situation and their stake in those properties. In particular, property owners ought to know that their land rights are yet to be framed let alone be granted and that the land around them is used with permission of the Client who still holds the title, much in the same way as a landlord. Property owners have therefore no right to undertake any alterations to the land and other developments like tuck shops and other business extensions that do not constitute part of the original buildings could end up in someone else's land. In their own interests they are also to be made aware that access to the main road could change in the course of the cadastral survey, since each plot must have a private access to public right of way.

TIME SCHEDULE AT INCEPTION

The Director General
Tanzania Ports Authority
P. O. Box 9184,
Dar Es Salaam,
TANZANIA.

Attention: DETS/Chief Estates & Rating Surveyor

12th September 2005

Your Ref: EN/5/4/2
Our Ref: TC/TPA/3

Dear Sir,

RE: CADASTRAL SURVEYING OF FORMER TPA HOUSING ESTATES
AT DSM, TANGA, MTWARA

SUBJECT: Tentative Time-Schedule at Inception

Please find herewith our second tentative time schedule to guide the flow of events on this project. The schedule is now dated to commence on 12th September 2005 and runs through to the expected day of completion, six months later, which end on Saturday, 11th March 2006.

The tome schedule is labelled tentative at Inception as it could change after visits to the respective sites and probably after submitting the inception report to client.

Kind Regards,

Dr. F. N. Lugoe

Managing Director

CADASTRAL SURVEYING OF FORMER TPA HOUSING ESTATES AT DAR ES SALAAM, TANGA & MTWARA

Schedule Commences on 12th October 2005 and ends on 11th April 2006

NOTE: Mobilisation/Demobilisation includes time for maintenance of equipment & vehicles as well as local information gathering.

THE INCEPTION REPORT

This report includes an attachment of High Resolution Ortho-photo Maps for all Dar Es Salaam Estates and

an Appendix of Subdivision Plans for all Estates in
All Three Sites

1. PROJECT HANDOVER/TAKEOVER: REPORT ON INSPECTION OF TANZANIA PORT AUTHORITY'S (TPA) HOUSING ESTATES IN READINESS FOR CADASTRAL SURVEYING FOR INDIVIDUAL TITLING

1.1 Introduction:

The tender documents for the Cadastral Surveying of Former TPA Housing Estates define the task ahead as one that aims at enabling each property owner to be issued with a letter of offer of a right of occupancy and later on, a certificate of occupancy.

To this effect a contract was signed on 31st August 2005 between TPA, the Client and Topo-Carto Consultants (T-CC) Ltd, the Consultant. The handing over of the work sites to the Consultant and inspection of the former houses of the Tanzania Ports Authority (TPA) started on 12th September 2005. It was conducted by a team of the Consultant in the company of TPA's Chief Estates and Rating Surveyor Mr. J. M. O. Safari. The reconnaissance and inspection had the following objectives:

- To hand over and take over the assignment as per contract agreement between TPA and Topo-Carto Consultants Ltd.
- To receive and confirm some of the problem areas requiring special attention of the Consultant.
- To appreciate the magnitude of the task embedded in the agreement for the cadastral surveying of former TPA housing estates at Dar es Salaam, Tanga and Mtwara.

The exercise was accomplished within two weeks although the actual time spent in the field was shorter but, the physical separation of the sites involved long travel by road and air. The Dar es Salaam site was handed over on Monday 12th September 2005. Tanga was handed over on Thursday 15th and Friday 16th September 2005. Mtwara was handed over on Friday 23rd September 2005.

1.2 Major Visit Revelations:

A visit to all sites has revealed the following situations:

1. Estates that are a subject of this exercise contain more than residential properties. In other words, there are properties on these estates that have not been disposed away and remain the properties of TPA or Tanzania Buildings Agency, TBA. Among these landed properties are; nursery schools, social clubs, sewers and sewerage plants (existing or planned), signal towers, play grounds, etc.

2. Following from (1) above also is that TPA seems to have no other options save to survey all properties and the open spaces in its estates if individualized title for buyers of the Ex-TPA stock of houses are to be obtained. The practice in titling is that each land parcel (urban plot) must have access to a public right of way but currently, all streets within an estate are private to TPA who owns the title to that estate. These will only become public either upon cancellation of the registered survey plan or upon a resurvey that re-defines the road/street network or both

3. The buildings identified and constituting the essence of subdivision surveys in the contract agreement generally fall into five categories and accordingly, the cadastral

survey will be specific to each of these categories. These categories are:

- Single houses on a housing estate,
- Single storey buildings with double occupancy either on a housing estate or stand alone plots,
- Single buildings of a maisonette type (on two floors) consisting of more than one unit and located on a housing estate or a stand alone plot,
- Blocks of flats of two or more storeys with common staircase and shared corridors,
- Two storey buildings of two occupants and one occupant on each of the two floors.

2. VISIT REVELATIONS PER SITE AND ESTATE

The following is a summary of major observations by the Consultant for each of the estates. Given in the summary are also decisions made by the Consultant subsequent to the visits aimed at facilitating the subdivisions or otherwise. The observations are also made in light of the data and information sources available up to the time of this report, in the form of maps plans etc,.

DAR ES SALAAM:

2.1 MSASANI ESTATES 1

- The houses of this estate are situated between Chole and Slipway Roads.
- There are 16 single buildings with single occupancy, arranged in a rectangular fashion around an open space The available engineer's layout plan shows labels the open space as a green zone. A designed access to the open space is blocked by house no. 12 (see layout). There is also a small concrete kiosk and two containers in this open space, in front of house number 13.
- There are also Semi detached maisonettes forming a U-arrangement around the single houses but separated by internal estate road. The houses have outlets to both the internal access and the outer public roads.
- Available documentation include the engineer's layout plan, a 1: 2,500 topographical map, and a recent high resolution orthophoto map These documents reveal that there are unplanned openings into the green zone and other minor nuances on the estate.

- It is the Consultant's position that the engineer's layout prevails in the plot layout and cadastral survey of the estate, be followed supplemented by the two maps of the area.
- There is no survey plan of the estate but ortho-photo and Township maps for the area to guide plot demarcation.
- The Consultant has therefore designed the subdivision layout based on available maps and plans. The semi-detached maisonettes will be located on a singe plot for each until such a time that the issue of partitioning along common walls has been resolved with the Town planning authorities at the MLHSD (see section 4 on huddles).

2.2 MSASANI HOUSING ESTATE 2.

- The housing estate is situated between Chole road on the west, Kahama road on the south and DANIDA housing estate to the east.
- There is no engineer's layout plan provided for the estate. However, it is well depicted on the township map sheet of Msasani and on the recent high resolution and colour orthophoto maps

- The estate is known as plot number 1826 and has a certificate of occupancy number 396274 of 1st January 1991 covering an area of 10.4 hectares. It is also engulfed by survey plan no D'827/8 to the North and Plan E'210/26 & E'210/27 to the South.

- It has been noted that there are extensive developments on this estate involving building extensions, wall building, and fencing accompanied by blocking of the side drive road to the north, west and south. Extensions on House no. 112 (see layout plan) are being made into the land set aside for the high-tension power line and common septic tanks. It has also been noted that some houses on the eastern side have no direct access to the public roads whilst two roads around the nursery school have been closed.

- The Consultant has subsequently prepared a plot layout based on the township map, the colour orthophoto map and ground reality. In this plan a road has been considered on the eastern side starting from house number 112 southwards for the benefits of houses number 112, 81, 82 and 90 which are other wise inaccessible without criminal trespass once individualized title is available.

2.3 PLOT NO 95, GUINEA ROAD, OSTERBAY.

- The plot has two buildings namely a main building with double occupancy (semi- detached) and a servant's quarter/garage to the north. The survey plan for this plot is yet to be found.
- The intention is to subdivide the plot into two should it be allowed by the Human Settlement Development Division, HSDD of the Ministry of Lands and Human Settlement Development, MLHSD (cf. section 4 on huddles). The subdivision will place the servants quarter/garage into the ownership of one of the two occupants and not both, because of its location (see orthophoto map)
- There is a large-scale topographical map and orthophoto map for this plot to guide into the subdivision process.
- The Consultant has prepared a subdivision along the partitioning wall of the two houses. One occupant will have no option but to surrender the servant's quarter/garage to the other. This

occupant enjoys however a larger share of the plot than the one who will retain the second building.

2.4 PLOT ON KIMWERI ROAD, OSTERBAY

- The plot has no survey plan, no plot number and no engineer's layout has been provided to the Consultant. The area has been identified on the high-resolution orthophoto map of Dar es Salaam at the corner of Kimweri and Kwale roads in Oysterbay.
- The building consists of four adjoined maisonettes that can easily be partitioned along the common walls and be subdivided into four plots as each occupies bottom and upper floor, should this be allowed by town planning authorities.
- The Consultant has subsequently prepared a subdivision layout that considers a cadastral survey process of this area into four plots as described above.

2.5 PLOT NO 2114/5, BLOCK A, SEA-VIEW, UPANGA

- There are three blocks of flats of six occupants each on this plot. In each block of flats the occupants share a common staircase, the roof, wastewater drainage system and access road. Under provisions of the laws in Tanzania (see section 4.2 on condominiums), a subdivision survey of this plot will be limited to providing land for each block of flats. In other words, the plot will be divided into three plots, with a view of providing common ownership of land rights to all occupants of each block of flats.
- There is a registered survey plan, a large-scale topographical map and high-resolution orthophoto map for this plot but there is no engineer's layout plan. The available documents are handy and sufficient in designing a layout plan.
- The Consultant has prepared a layout plan for the subdivision of the plot into four using the maps available and the ground truth.

2.6 PLOT NO 2104/5, OCEAN ROAD, UPANGA

- There are two flats of five storeys each on this plot. The larger block is located along the Ocean Road. The smaller block, has half the number of flats as the larger will be facing the nearby street or be provided with an easement if it is to be surveyed in own plot. There is a large-scale topographical map, high-resolution Orthophoto map, but there is no registered survey plan and no engineer's layout plan.

- Having considered all these factors and the ground truth, the Consultant has designed a layout for the subdivision of this plot into two plots aimed at common ownership of the separate plots by the owners of the flats in the building for each plot.

2.7 PLOT NO 269, ALYKHAN ROAD, UPANGA

- The building on this plot is a double storey with double occupancy. Each occupies top and ground floor divided by a common wall along the east-west direction.
- The only information available for this plot is the high-resolution orthophoto map. There is no registered survey plan, township map, and no engineer's layout plan either.

- The Consultant has prepared a subdivision plan for this plot bearing in mind the ground situation and the aerial image as provided by the orthophoto map.

2.8 PLOT NO 272, ALYKHAN ROAD, UPANGA

- The plot contains a double storey house with double occupancy whereby one occupies the ground floor and the other one occupies the top floor. There is also a grocery in front of the block but does not form part of the main building.
- The Consultant proposes that the existing certificate of occupancy be transferred to the co-occupants as it is, since subdivision is not possible.

2.9 PLOT NO 194, MALIK / UNDALI ROAD, UPANGA

- The building on this plot is two storey. It is occupied by four residents - two on the ground floor and two on the upper. Top floor residents share common staircase and corridor
- There is no registered survey plan and no engineer's layout plan for this plot. There

is however an orthophoto map and the plot is displayed on the Township map at a 1: 2,500 scale.

- It is not possible to subdivide the plot and facilitate individual titles on this plot. The Consultant therefore proposes that the existing title be transferred to all the owners in common ownership.

2.10 PLOT NO 577, MINDU STRET, UPANGA

- On this plot is located one double storey building of eight occupants. Each building unit is a maisonette type in which each owner occupies both ground and upper floor. There is enough land on the south side to provide an easement and hence private access to each unit thus changing the main entrance from the northern to the southern side.
- Land information available for this plot includes topographical large-scale map and an orthophoto map in full colour. There is however no survey plan and no engineer's layout plan.
- Permission will be required for the MLHSD to accept the common walls as constituting plot boundary (see section 4)

and proceed with the subdivisions along the wall lines and thus provide 8 plots for individual titling.

- The Consultant has prepared the subdivision layout along the above suggestions pending the granting of such permission. The layout is made using the enlarged topographical map of the area coupled with ground and other information from recent orthophoto maps of the area.

2.11 PLOT NO 569, MINDU STREET, UPANGA.

- On this plot, the main building is a double storey building with double occupancy. The upper floor occupant shares the ground with the ground floor occupant. There is no possibility of granting private access to either occupant and hence it is not possible to provide individual titles on this plot.
- Data for the plot is available in the form of an orthophoto and topographical maps. There is no survey plan and no engineer's layout, yet available.
- The Consultant advises that in view of the aforesaid it is not possible to

subdivide the plot. It is proposed: that common ownership of this plot be granted to both occupants.

2.12   PLOT NO 446, CHARAMBE STREET, UPANGA

- The building here is a block of six flats on three floors sharing a common staircase
- Data available here, includes topographical & orthophoto map sources but there is neither engineer's layout nor a registered survey plan etc.
- It is not possible under current practice in Tanzania to subdivide this plot It is therefore proposed that all owners be granted common ownership of the plot and its landed property.

2.13   PLOT NO 454 CHARAMBE STREET – UPANGA.

- On this plot is a three-storey building with three occupants - one occupant in each floor.
- Land information is available in the form of an orthophoto map only. There is no

topographical map, no registered survey plan and no engineer's layout plan.

- This case is similar to that of Plot 446 on the same street. It is proposed that the three occupants be granted with common ownership of the landed property since vertical subdivision cannot be effected.

2.14 PLOT NO 723, UPANGA

- This property could not be located.

2.15 PLOT NO 2445/208, MSIMBAZI

- This plot is situated between Msimbazi Street to the east, Lindi Street to the north, a railways estate to the west and the central railway line on the south. It has a general survey plan No D 248/4 for a plot covering an area of 3.6206 hectares. The estate is encumbered with tuck shops, vehicle garages and other petty businesses on all sides and on the estate itself. The open space on the south side is an open-for-all playground. The worst situation of encumbrances can be witnessed along Msimbazi and Lindi Streets.

- Land information here is scanty, but a recent orthophoto map and a layout plan for the houses are available to guide and determine the demarcation of plot boundaries. The estate consists of 106 semidetached houses and can be surveyed into equally many co-occupied plots of two occupants each until such a time that permission is granted to partition the plots along the shared walls (see section 4).
- The Consultant has designed the subdivision layout bearing in mind the above situations.

2.16   MGULANI HOUSING ESTATE

- The Mgulani housing estate is located at the corner of Nelson Mandela and Kilwa roads behind the Fuelling Station. It extends northward to Kurasini Road leaving out the corner of Kurasini and Kilwa roads The northern part consists of blocks of flats whilst the southern is made up entirely of single-storey buildings of mixed single and double occupancy.

- The land data on this estate is partial as parts of the engineer's layout are missing and one sheet of the 1:2,500 for the area is out of stock and the registered survey plan is yet to be found. However, the orthophoto map for the estate gives a nice display of the reality on the ground.
- Along Kurasini road is a cluster of flats of four residents, two on top floor and two on ground floor. The blocks cannot be partitioned and the Consultant therefore proposes that common ownership for each block be considered.
- Most single-storey houses of single occupancy have direct access to the road. There are also semi-detached houses of two occupants but one has access only through the frontage of the other. The plot design must rectify this situation. There is also a mosque along the estate boundary to the east but without direct public access. This will be excised from the estate.
- The Consultant has designed the subdivision layout considering the ground truth from the field visits, existing orthophoto and topographical township maps of the area and the engineer's layout plans

2.17 KURASINI HOUSING ESTATE, PLOT NO 219 – 307

- Kurasini housing estate is located to the south of the TPA headquarters and is comprised of a mixed category of buildings including single storey detached and semi-detached houses and blocks of flats. The estate covers 17.71 hectares of land and in addition to the dwellings has a club house, shop, primary school, waste water treatment plant and a large open space
- Land information is provided by a registered survey plan no. 21081, a certificate of title no. 40068, township topographical map at 1: 2,500 scale, recent high-resolution orthophoto map, and engineer's layout plan of the buildings and infrastructure. A coordinate list of the boundary beacons, in Dar Local System, is also available and has been transformed by the Consultant to UTM in Arc 1960 Datum. It has been noted that some of the old buildings are not shown on the engineer's layout plan and many of these

have no direct access to the public right of way.

- Land information described above and ground truth have guided and determined the design of a subdivision layout by the Consultant. But, as in previous cases, semi-detached buildings will be located in a single plot for each building until authority is obtained for their partitioning. The two blocks of flats are dwellings that can only be occupied in common (co-occupancy).

2.18   KIGAMBONI ESTATE:

- Kigamboni housing estate is located on the Kigamboni beach and is bordered by the Kigamboni Ferry Station on the west. It is a big estate that is only partially developed and partially neglected. The estate hosts the Kigamboni Signal Station, which is a vital installation in ship navigation through the Dar Es Salaam Harbour. There is also a nursery school on this estate. All houses on the estate are semi detached and of double occupancy each.
- Land information on this estate is provided by the township large-scale

topographical mapand the high-resolution orthophoto map. The estate is provided with an engineer's layout plan. The survey plan and corresponding data are yet to be found.

- The Consultant has designed the subdivision layout plan for the estate that also introduces streets to facilitate private access to public right of way for each owner. It is also envisaged that each plot will initially provide for one building until such a time that partitioning of the plots along the shared walls will be permissible.

2.19   CHANG'OMBE BLOCK 'C', PLOT NO 1-10

- Located along Chang'ombe road is this estate of five semi-detached buildings of double occupancy each. All houses are similar in construction.
- Land information is provided by the township map at 1:2,500 scale, a high-resolution orthophoto map in full colour, and a registered survey plan whose data is yet to be found. The absence of an engineer's layout plan will be

supplemented by available data and the ground inspection reality.

- The Consultant has developed the subdivision plan for the estate that provides five plots for each building. The plots will be partitioned after authorities have granted relevant permission.

2.20   KILWA ROAD, PLOT NOS. 43 & 44

- The plot is situated at the corner of Kilwa and Bandari roads – a place known as Mivinjeni. The plot has three Blocks of flats of two storeys each, and a servant's quarter. There are 16 terraced flats and eight servant quarters on the estate.
- Land information here, is given on the township topographical map, the orthophoto map of Dar es Salaam, and engineer's layout to guide the physical positioning of the plots and a registered survey plan is also available.
- The Consultant has prepared the subdivision plan for the estate that will provide individual plots for each block of flats aimed at granting a common title to the occupiers of each building.

2.21   GEREZANI HOUSING ESTATE

- The estate is situated at the roundabout, at the corner of Gerezani Street and Bandari road. It is located within Dar-es-salaam Port boundary. The survey is meant therefore not only to provide for ownership of these properties but also to excise the land from that of the Port. These are two single units numbered G38, and G39, of single occupancy each. There are extensive developments taking place at both houses that go far beyond the fringes of the dwellings into the bounds of a 5 m high depression to the port basin. The Consultant will ignore these developments and provide land for the existing dwellings only.
- Land information is given by the Dar Es Salaam orthophoto map and the Dar Es Salaam township map for the location and demarcation of the area.
- The Consultant has designed the plan for the excision and survey of this estate into single independent land parcels.

TANGA:

2.22   HOUSE NO. 114, FIRE BRIGADE

- The plot is located near the port and it is comprised of a two-storey building with four occupants on the ground and four on the upper floors. The flats share a common staircase and entrance.
- Plot location by a hand held global positioning system (HHGPS) for identification on a township map is (9440164 N & 511628 E). The registered survey plan for this plot, if at all it exists, is yet to be found.
- The plot cannot be subdivided to give individual title to each occupant. The Consultant therefore proposes that the existing title be transferred to all eight in co-occupancy.

2.23   USAGARA ESTATE

There are two estates standing side by side namely;
- Block E, F, G, & H (Usagara Estate)
- Block J (New Usagara Estate)

- Land information on both new and old Usagara Estates is given as deed plans obtained from the certificates of occupancy, a township map and an engineer's layout plan to guide plot subdivision. Usagara estate is comprised of Plot nos 14-26, 10-18, 6-11 and 1-22 on blocks E, F, G & H:

- This plot is comprised of 23 semi-detached houses of variable occupancy ranging from single to five occupants. There is also an adjoining house, excluding house no 90, a shop, a nursery school and open spaces.

- The Consultants subdivision plan is presented herewith. The estate was handed over but <u>does not appear in the tender documents</u>.

2.24   NEW USAGARA ESTATE

The estate is comprised of Plot nos 20-56, 58, 60, 62, 64, 66, 67-93 and plots 104 – 107 in Block J:

- New Usagara Estate was not listed in the tender documents and hence was not included in the contractual agreement. The Client is now desirous that this estate form part of the cadastral survey plan.

- It comprises of three four-storey blocks of flats known as blocks A, Band C with 16,16 and 8 units each respectively.

Further, there are two adjoined blocks of flats at (9438978N, 512559E) with 48 flats, two houses AH and BH at (9439210N, 512265E), a social hall, four single units numbered D, E, F, & G, a police armory and a large open space across the street from the three blocks of flats. All flats have common staircases and corridors.

- There is no engineer's layout plan but a topographical large scale map showing the location of all buildings. The Consultant has utilized this data to produce a subdivision layout plan.

2.25   CHUDA BLOCK – Flat Nos. 61-63

- This estate is locatable by H.H.GPS at (9439080 N & 510484 E), KANA. It is comprised of Blocks of Flats with two occupants in each. There is a public road on the East side of the estate. Chuda Plots 60 and 61 do not appear in the tender documents.
- Subdivision surveys can be described within the same scenario as for the Fire Brigade estate.
- Land information is obtainable from the deed plan and township topographical

map, but there is no engineer's layout plan.

- The Consultant has developed a subdivision layout plan guided by topographical maps and ground inspection reality.

2.26   CHUDA House No. 65

This estate can be located by HHGPS at the following co-ordinates: (9439142 N, 510418 E). It is a block of flats of three storeys high and four residents on each level for a total of 12 occupants. The flats share common staircases & entrances.

Land information is provided by a topographical township map (1:2,500). There is no registered survey plan and no engineer's layout plan provided as yet. A subdivision plan has been prepared by the Consultant in readiness for the cadastral survey.

2.27   CHUDA House No. 29

- The estate is locatable by HHGPS at: 9439254 N & 510370 E. It constitutes of two houses with five occupants in each and a third one co-habited by two on the opposite street.

- Land information is given as a deed plan for plot demarcation; a township topographical map and a registered survey plan to determine the plot boundary and subdivision plan. Please see relevant plans.

2.28   CHUDA House No. 30

- The estate is locatable by HHGPS at: 9439268 N & 510292 E, near railway sliding. It is comprised of three single houses of two residents each to a total of six occupants on this plot.
- Land information is obtainable on the plan no D' 192/4 with registered plan no 22/06 covering an area of 31885sq meters, a coordinates list for recovery of the missing beacons, and a township map to determine the location area. There is no engineer's layout plan.
- Subdivision plan has been prepared to enable titling.

2.29   CHUMBAGENI, - Block H, Plot No 94

- The plot is locatable by HHGPS at: 9439880 N & 510039 E with single houses on a common title and which are to be subdivided for provision of individualized titles. The plot lies along Chumbageni road and Jamhuri ring road with 22 housing units. The survey plan and engineer's layout plan are yet to be provided.
- Land Information on the estate is available on a township topographical map to guide the subdivision of plots, a deed plan to guide recovery of plot boundary.
- The Consultant has prepared a subdivision layout based on available land information and land inspection reality.
- This plot <u>was not included in the tender documents</u> but the Client is desirous that it forms part of the current exercise

2.30 CHUMBAGENI Block KB VII, PLOTS NO 54, 55 & 56

- There is a single storey building in each of the three plots. Other detail is similar to plot no 94 of block H above but no subdivision will be required as each house occupies one plot. This estate <u>does not appear in the tender documents.</u>

- This is a direct cadastral survey of each of the houses in the estate.

2.31   CHUMBAGENI Block KB VIII, PLOTS NOS. 129, 131, 133, 137 – 147

- The plots are locatable with aid of a HHGPS at: 9439530 N & 509427 E. These are single detached houses with each plot occupied by one resident to a total of 15 occupants.
- The house of PLOT NO 135 was burnt by fire and later taken away from the TPA stock, i.e., it no longer belongs to T.P.A. This will be omitted from the survey.
- Land information is obtainable on large-scale township map to determine the location area, deed plan to guide re-establishing the plot boundary. This information has enabled the consultant to prepare a subdivision layout, which is now available.

2.32   NEW NGUVUMALI –BLOCK 'B', PLOTS 27, 29, 30, 31 & 32

- This seems to be the same as Chumbageni Housing Estate, Block 'H', with similar numbering for plot numbers namely, Nos. 27, 29, 30, 31 and 32. Client to please verify or otherwise.
- The estate is locatable by HHGPS at: 9439214 N & 508794 E and on the township map which will guide the subdivision of plots. The deed plan will guide re-establishment of the plot boundary and the coordinate list for replacement of the missing beacons and survey datum.
- There are five houses on a common title, but each house has been built on a separate plot. Subdivision layout has been prepared and included in this report.

2.33   KISOSORA Block 'N', PLOT NOS. 5, 7 ,9 ,11 & 13

- This plot is locatable by HHGPS at: 9439970 N & 509371E where 3 houses of semi-detached type on five plots shall be subdivided into three plots initially and eventually to as many as the number of occupants.

- A subdivision plan is available in readiness for cadastral surveying as envisaged.

2.34   INDEPENDENCE AVENUE PLOT

- The plot is locatable by HHGPS at: 9439862 N & 511567 E. opposite NASSACO and Bandari buildings. The building hosts a dispensary on the ground floor but has two residents on top floor who require certificates of occupancy for their properties. The residents have separate stairs case with the western side resident using the hospital entrance.
- There is no survey plan, Layout and topographical map obtained as yet.
- The plot was not provided in the tender documents but the Client finds it necessary that it forms part of this cadastral surveying exercise.
- The plot <u>does not appear in the tender documents</u>.

MTWARA:

2.35   PLOT 242 LIGULA

This is a block of two flats locatable on a map by HHGPS coordinates at (8863856N, 0628222E). Each resident occupies own floor and share the garage/car park. The property cannot be partitioned into two independent pieces of land. The Consultant therefore advises that a common title be transferred to the two occupants

2.36   RAILWAYS ESTATE

The Estate is locatable by HHGPS coordinates (8864822N, 630493E). It is located close to the Tanzania Ports Authority headquarters in the municipality of Mtwara starting from the corner of Port and Makonde Roads. It is bordered by Shangani Primary School on the west and the Railways quarters on the east.

Only seven houses on this estate are of single occupancy. The remaining are single-storey houses of double occupancy. The houses share pit latrines and bathrooms which are located some distance from the main houses. The estate hosts a nursery school and club. There is a large open space that could come handy in future development.

Land information is given on the uncompleted topographical sheets of 1:2,500 of the municipality. A copy of the relevant draft copy has been obtained from the Surveys and Mapping Division in Dar Es Salaam. The registered survey plan no. 27505 is yet to be obtained. However, there is a copy of the deed plan for use in proper identification of the estate boundary and an elaborate engineer's layout plan of the estate.

The Consultant has designed a plot subdivision plan for all houses in focus. The survey will proceed with the understanding that each house will be provided with land and private access. In view of the prevailing uncertainty in subdividing a plot along a shared wall, this subdivision will only be undertaken after the authorities have assented to it (see section 4).

2.37   Post-Handover Work Summary:

Estate	No. of Buildings or Plots*	No. of Occupants	No of Plots After Final Subdivision	Other Plots**
	DAR ES SALAAM SITE			
Msasani Estate 1	23	30	30	1 green area
Msasani Estate 2	65	83	83	1

				nursery 3 open space
Plot No. 95, Guinea Rd, Oysterbay	1	2	2	
Plot on Kimweri Rd, Oysterbay	1	4	4	
Plot No. 2114/5 Block-A Seaview, Upanga	3	18	3	
Plot No. 2104/5 Ocean Rd, Upanga	2	15	2	
Plot No. 269, AlyKhan Rd, Upanga	1	2	2	
Plot No. 272, AlyKhan Rd, Upanga	1	2	1	
Plot No.194, Malik/Undali Road, Upanga	1	4	1	
Plot No. 577, Mindu Str., Upanga	1	8	8	
Plot No. 569, Mindu Str., Upanga	1	2	1	
Plot No. 446, Charambe Str., Upanga	1	6	1	
Plot No. 454, Charambe Str., Upanga	1	3	1	
Plot No. 723, Upanga	?	?	?	
Plot No. 2445/208 Msimbazi	53	106	106	1 open space
Mgulani Housing Estate	111	275	158	3 open space 1

				mosque
Kurasini Housing Estate Plot No. 219-307	90	150	140	1 T/plant, 1 shop 1 P/School 1SocialClub 4 Open spaces
Kigamboni Estate	5	10	10	1 signal station 1 nursery 2 open space
Chang'ombe Block 'C' Plots 1-10	5	10	10	
Kilwa Road, Plots No. 43 & 44	3	16	3	
Gerezani Housing Estate	2	2	2	
Sub Total 1	371	748	568	
		TANGA SITE		
House No. 114, Fire Brigade	1	8	1	
***Usagara Estate, Plots 14-26, 10-18, 6-	23	90	90	1 shop,

11, 1-22 Blocks E, F, G & H Resp.				1 nursery 1 open space
New Usagara Estate, Block J, Plots 20-56, 58, 60, 62, 64, 66, 67-93 & 104-107	10	94	10	1 social hall 1 P/arm ory 1 open space
Chuda/Kana Block Flats 60-63	3	6	6	
Chuda House No.65	1	12	1	
Chuda House No. 29	3	12	4	
Chuda House No. 30	3	6	6	
***Chumbageni Block H, Plot No. 94	22	22	22	
Chumbageni BlockKB VII, Plots Nos 54-56.	3	3	3	
Chumbageni Block KB VIII, Plots No. 129, 131, 133, 137, 138-147	15	15	15	
New Nguvumali Block 'B', Plots No. 27, 29, 30, 31 & 32.	5	5	5	
Sub Total 2	89	273	163	
MTWARA				
242 Ligula Flat	1	2	1	
Railways Estate	91	175	175	1 nurser

				y 1 social club 1 mosque, 1 shop 6 open space
Sub Total 3	92	177	176	
Grand Total	**551**	**1196**	**913**	

* The prevailing practice is to fit one unique building in one plot

**Client to advise whether to survey as single open spaces or subdivide into plots

***(90+22) Plots, etc., were not included in the tender documents

3. POST-HANDOVER PREPARATIONS

3.1 Survey Instructions:

The Project Agreement documentation calls for the Consultant to deliver the services in accordance with Survey Instructions of the Commissioner for Lands. Licensed surveyors are required by law to, additionally, obtain survey instructions of the Director of surveys and mapping. The Consultant has accordingly applied for the instructions from the Director of the Surveys and Mapping Division, pending the Client's application and grant of permission to divide the land.

It is worthy of note that the survey instructions of the Commissioner have traditionally, been preceded by an application for permission to divide the land. At the same time the survey instructions of the Director of SMD await those of the Commissioner for Lands. Where problems are not anticipated surveyors have often proceeded with surveys pending the issuance of survey instructions, by both authorities. The Consultant is therefore proceeding with the project, where these are not hindered by the absence of survey instructions.

The Client is therefore requested to offer assurance that the project will not be hindered by a denial of survey instructions for this project emanating from a reluctance or otherwise of the CoL to grant permission to divide the various estates which are under title.

3.2 Permission to Divide Land under Title:

This consideration emanates from a provision of section 83 of the Land Registration Ordinance (Cap 334) that has provided official form L. R. 30, being an application for the "Division of Registered Land".

Such an application must contain an attachment, being a plan of the subdivision. The Consultant has prepared such plan for each of the estates based on a combination of large scale (1:2,500) township topographical maps, recent high-resolution orthophoto maps for the Dar es Salaam estates, ground reality as evidenced by recent inspection of the estates and environs, and engineer's layout plans for the development of the estates where available.

In using the large scale topographical maps, the maps were first enlarged to provide a clearer picture of the arrangement of the buildings and infrastructure as made by the engineers and as existing at the time of preparing the maps. Any recent developments were captured on orthophoto maps. In some instances, lay out plans of the engineer/developer exist and these have also been used in the preparation of subdivision plans and will be used in the staking out of plots during cadastral survey field work.

There are however, estates for which engineer's lay out plans either original or 'as made layouts' are vital to this exercise but have not been handed over to the Consultant. The Client is requested to expedite their delivery to the Consultant. The estates are: Msasani Housing Estate 2, and Mgulani Housing Estate in Dar Es Salaam; and Usagara Estate, New Usagara Estate, Plot 94 Chumbageni Block H, Chumbageni Block KB VIII and New Nguvumali Block B in Tanga.

In addition to the use of these documents (layouts, maps and plans) in guiding the determination and survey of plots, the layout and plans will also assist in the application for permission to subdivide the estates. But, as the Land Registration Ordinance (section 83) states, it is an application of the 'owner of the right of occupancy' and ought to be made therefore by the Client. The plans are provided herewith to assist the client in lodging applications for permit to subdivide the various estates. Also provided in this documentation are the relevant orthophoto maps for Dar Es Salaam estates.

3.3 Consequences of the Permission to Subdivide the Estates:

Should the Commissioner accept the application and grant permission to subdivide the estates, as it is expected, the **original titles will be revoked and the corresponding registered survey plans will be cancelled** paving way for new survey plans to be prepared and registered so as to guide the issuance of new individualized certificates of occupancy for each plot. The attention of the Client is drawn to the fact that the cancellation of existing documents could lead to: the Client losing some of the land he still needs and to mismanagement of open spaces. It is of note that:

- Estates that are a subject of this exercise contain more than residential properties. In other words there are properties on these estates that have not been disposed away and remain the properties of TPA or Tanzania Buildings Agency, TBA. Among these landed properties are; nursery schools, social clubs, sewers and sewerage plants (existing or planned), signal towers, play grounds, etc. The Consultant needs clear guidance as to what should be done with the land not forming part of residential houses but containing other TPA interests on the estates.

- There are also open spaces on several estates, which were either meant for use by residents on the estates or for future development and use by the Client. Cancellation of the documents will mean that any vacant land on the estates will revert to the planning authority, which in this case is either the Municipality or the City Council. Consequently, the Client will lose land, which could have been used for future development. An example is the possible land requirements for the Southern Development Corridor in Mtwara, etc.

But besides the would be needs of the Client, experience has shown that immediately after it becomes public knowledge that TPA has pulled out of the open spaces, there will be land-grabbing worse than what has happened to the Msimbazi Estate, so far. It is also of note that since the vacant is unplanned, then any subsequent development by land grabbers will deteriorate into squatter land and the

Client will be blamed for not being visionary enough as to prevent such a situation from occurring at a time when the Government is spending huge sums of money on regularization of titles (Sub-Part 2 of Part VII of the Land Act No.4 of 1999). A direct consequence of this development to the TPA workers is that their properties will lose value and could lead to court cases with compensation claims.

- Proper land disposition is one that finds a legal owner as and when the Client's land interests in the estates are extinguished. In this case all land on the estates including open spaces and non-residential properties must find legal owners. This is possible if all the land is surveyed at the time of subdividing the estates for transfer of ownership of residential properties to TPA workers. Such an approach will redefine the public rights of way that are not encumbered by TPAs original land title and will vest the rest of the non-residential properties to self, pending another phase of disposition should that be in the interest of the Client. As a suggestion, the large open spaces may be surveyed into additional residential plots to be allocated to Client's staff who may be landless but were not offered any properties during the recent sales. of the Client's stock.

4. TITLE INDIVIDUALISATION HUDDLES:

4.1 Land Parcels of Semi-detached Houses and Adjoined Apartments:

Many houses listed for cadastral surveying in this project constitute semi-detached dwellings in which a wall is shared between two adjacent neighbours. The Town Planning Section of the Human Settlement Development Division (HSDD) has always insisted that a land parcel (or plot of land) must be an area of land under homogeneous property rights and unique ownership. Accordingly, where a wall is shared, the land parcel ceases to provide that homogeneity and uniqueness. In the absence of a Condominium or Strata Titles Act in Tanzania, the Town Planning authorities do not recognize a shared wall to be part of a plot boundary. Consequently, semi-detached houses and adjoined apartments can only be occupied in common as per Part XII of the Land Act No. 4 of 1999.

The Town Planning description leans more towards building regulations and building permits in urban areas and not on boundary delineation and demarcation. On the other hand, and from the cadastral survey point of view, land parcel or plot boundaries are defined by thin lines between centres of corner beacons or other markers and, can be defined irrespective of the envisaged property development or mode of occupancy of adjacent land parcels. However, the purpose of conducting a cadastral survey in a subdivision is to eventually obtain a land title for the land parcel in question and it is particularly for this reason that the position of the HSDD be seriously considered. As of now, the Client is assured of cadastral surveying of plots for individual buildings but not the partitioning thereof unless a consensus is arrived at on the matter.

The above advice must be given serious consideration if individualized titles are to be obtained for all those units in shared buildings. It is further noted that the majority of houses and blocks of flats belonging to the Tanzania Buildings Agency and probably the National Housing Corporation will benefit immensely from such a provision.

In the absence of an agreement on this issue subdivision will follow the current practice and the current legal provisions. The plots will be surveyed for each building and titles in such buildings will be titles in common. The Consultants will proceed with survey in this context pending the Clients Instructions to the Contrary.

4.2 Titling in Condominiums:

It is currently not legal to grant individualized titles in shared buildings or condominiums. Co-occupancy of the joint or common type is the only avenue available to titling in shared accommodation. This shortfall is brought about because Tanzania does not have a legislation providing for strata titles. In such a case only co-occupation is possible and the land and property shall be occupied in common (see Part XII of the Land Act No. 4 of 1999) by all who have purchased property in a given unit of a vertically stratified block of flats and semi-detached houses.

It is the exclusive advice of the Consultant that in view of the likelihood of a Sectional Titles or Condominiums Act to be enacted in the foreseeable future, itself being provided for under section 6.1 of the 1995 Land Policy document, a representation be made to the Acting Director for HSDD and the Commissioner for Lands and a permission be sought for the two departments to accept a shared wall as part of a boundary between developed plots or land parcels. Such acceptance by the HSDD can pave way for an inclusion of a condition in the Certificates of Occupancy of the neighbors sharing a wall for its safeguard as a recognized boundary between their plots. In other words, it is proposed planning authorities be prevailed upon to relax their standards and specifications in this matter of national significance so that the subdivision of land be done along the walls/lines separating the different units in shared buildings. Such relaxation will facilitate grant of individualized titles in semi detached houses, which constitute the majority of shared accommodation in this project.

4.3 Awareness on Landed Property and Land Rights:

The reconnaissance tour of the Consultant in the company of the Client's Chief Estates and Rating Surveyor, Mr. J. M. O. Safari, has revealed that:

- Many occupants of the houses earmarked for cadastral surveying are busy with extensions, fencing, wall building and landscaping, among other activities to the land around their houses. In some instances, these developments have encroached on areas set aside for utilities while in other cases the side roads have been blocked altogether.

- The future of shared accommodation needs to be underscored as some neighbours are unaware that they are likely to be tied up in a common title for the foreseeable future.

- Many of the small houses have shared pit latrines located between houses, etc. It is prudent that they be informed of the real situation and their win or lose stake in those properties.

- In particular, property owners ought to know that their land rights are yet to be framed let alone be granted and that the land around them is used with permission of the Client who still holds the title, much in the same way as a landlord until an offer is made. Property owners have therefore no right to undertake any alterations to the land and other developments like tuck shops and other business extensions that do not constitute part of the original buildings could end up in someone else's irrecoverable domain.

- Finally, they are also to be made aware that access to the main road could change in the course of the cadastral survey, since each plot must have a private access to public right of way to prevent cases of criminal trespass on neighbors' properties.

A public education, awareness creation and enhancement needs to be undertaken by the Client along the above lines, but not limited to these, to facilitate peace and harmony among neighbors who are colleagues and prevent real estate squabbles from spilling over to the work place. Understanding the above points well will also facilitate trouble-free execution of this project by the Consultant.

5. MATTERS REQUIRING IMMEDIATE ATTENTION:

1. Provide Engineer's layout plans or the 'as made' drawings for the following estates and plots: Msasani Housing Estate 2, and Mgulani Housing Estate in Dar Es Salaam; and Usagara Estate, New Usagara Estate, Plot 94 Chumbageni Block H, Chumbageni Block KB VIII

and New Nguvumali Block B in Tanga. Please consult sections 2.2, 2.16, 2.23, 2.24, 2.30, 2.31, 2.32 and 2.33 for details.

2. An early decision on whether or not the project should be broadened to include all landed properties and open spaces, which will be affected by cancellation of the registered survey plan is being awaited. Please consult section 3.3 for details.

3. The estates handed over during the inspection tour, but were not included in the tender documents require an extension of time with pay. Please peruse through sections 2.23, 2.25, 2.30, 2.31 and 2.35 for details on this item.

4. Verify and confirm, or otherwise, whether or not New Nguvumali Block 'B' is the same real estate as Chumbageni Housing Estate since the tender documents and site identification seem to contradict.

5. As a matter of urgent attention the Client shall apply to the Commissioner for Lands for 'Permission to Divide Registered Land' and the Commissioner's survey instructions which go together (see section 3.2).

6. Conduct a dialogue with the HSDD's town planning authorities to settle the matter of subdivision of developed plots along shared walls. Please consult sections 4.1 and 4.2 for details.
7. Confirm, or otherwise, the data provided in the table in section 2.28 constituting the total workload in this project
8. Approve of the subdivision plans as provided in volume 2 of this report. Consult also section 2 on background data to these plans.

6. THE NEW TIME SCHEDULE

The new time schedule emanating from the inception activities is herewith presented. The sequencing of sites is could change easily as it depends on data availability particularly for Mtwara and the weather. As of now the Consultant has obtained data for some estates in Dar es Salaam and Tanga but not Mtwara and will therefore continue with Dar es Salaam until most of the data is available. Should the rains delay our preference is to complete Mtwara as soon as possible.

TOWN PLANNING DRAWINGS FOR THE CADASTRAL SURVEY OF FORMER TPA HOUSING ESTATES IN DSM, TANGA AND MTWARA

A tripartite meeting of TPA, TBA and the Consultant that met on the 11th November 2005 instructed the Consultant to engage a town planner and design layout plans of the estates for all lands on the estates that included the subdivision of the existing residential estates, the non-residential properties and the undeveloped areas. This short report covers the exercise and addresses issues emanating from the complete layouts. The Town Planning Drawing subdivisions of the various estates in this project have been completed.

All TP-Drawing subdivisions have been carefully designed within the existing and registered TP-Drawings of the respective areas as approved by the Director of Human Settlement Development (DHSD). The Drawings were informed by the subdivision plans submitted by the Consultant, in the appendix to the Inception Report, in addition to the large scale township maps and, of course. The town planner was availed with relevant sections of the inception report so as to get a feel of the existing situation on the ground. The layout designs were reviewed by the consultant so as to ensure the subdivisions did not unnecessarily depart from ground reality and expectations of the Client.

The layout plans for the subdivision of the estates show notable changes to the subdivisions proposed in the appendix to the inception report. These are briefly discussed hereunder:

1. Msasani Estate 2

The area TP drawing contained the subdivisions, which had not been adhered to at the time of developing this estate. The subdivisions this time have reverted to the previous general scheme after taking into consideration the developments on the site. Specifically

(a) The side roads alongside Chale and Kahama Road and diversions thereof have been removed

(b) An open space to the south of Houses 110 AND 111 has been maintained as open space for the high voltage power time and sceptic tanks. This has new called for a cul –de-sac to the south of MS 84 and 112 also, to give road access to MS 87 and 81.

(c) Another cul-de-sac has been introduced to provide road access to MS 82, 89 and 90. The open space to the opposite MS 89 has been left intract. It is the only open space on the estate and the client is advised to re-think the decision not to divide up the area into plots.

(d) As agreed in the tripartite meeting of 11th November 2005. The undeveloped areas have been transformed into plots. This includes the small strip of land at the North West Corner of the estate, which has been apportioned to Houses No. 70 and 79.

Summary: Number of developed plot: 85
New Undeveloped plots: 2
Open spaces: 1

2. Msimbazi Housing Estate Plot No. 2445/208:

The subdivision TP drawing for this estate has covered the undeveloped area with new plots. The developed plots have fully adhered to the proposals given in the inception report.

Summary: Number of Developed plots: 94
New undeveloped Plots: 18
Open spaces: 1

3. Kigamboni Housing Estate:

A close examination of this estate on the TP drawing for kigamboni furry Area has revealed that a modern dual carriage way has been designed along the same route as the existing road to Mji – Mwema. The new highway has a 45m reserve but extends more into the estate than otherwise. The consequence of this revelation is that three houses (of occupants) fall on the highway and the land cannot therefore be surveyed tilling. The client has no option but to remove the houses from the list of those earmarked for this exercise.

Summary: Number of Developed plots 6
New undeveloped Plots: 11
Open spaces: 0

4.   Plots No 43 + 44 Kurasini:

At the request of the Client design has been altered slightly so as to provide four plots.

Summary: Number of Developed plots: 4
New undeveloped Plots: 0
Open spaces: 0

5.   Plots No. 38 + 39 Gerezani:

The attention of the Client is drawn to a situation similar to that of Kigamboni Estate. The Gerezani

Street is earmarked for expansion. This expansion
will necessitate the demolition of both houses at
sometime. The land occupied by the houses is
therefore a road reserve and cannot be surveyed
for titling. Consequently, the Client. Must remove
the two from the list of the project.

Summary: Number of Developed plots: 0
     New undeveloped Plots: 0
     Open spaces: 0

6. Usagara Estate (Old + New):

The general TP drawing for usagara shows streets
on all sides of there estates, which have allowed
the subdivision to change from that proposed in
the inception report

Summary: Developed plots: 113
     New undeveloped Residential Plots: 34
     Open spaces: 2
     Play Ground: 1

7. CHUDA Plots No. 30

There are three houses of six occupants on this
estate. The TP drawing for Chuda / kana area
shows that the eastern most house was built
without regard for the railway reserve. This land

for this house cannot therefore be surveyed for titling and client must as a consequence remove the house from the project.

Summary: Number of Developed Residential plots: 2
Number of New Plots: 0
Number of open spaces: 0

8. Block E Chuda:

The estate has not been developed and is not occupied under any kind of Tenure. The plot has now been divided into 11 plots.

Summary: Number of New Residential Plots: 11.

Work summary:

The table below captures the work summary as quantified from the designed layouts of all estate. It displays the developed plots subdivided in accordance with the inception report recommendations and as approved by the Client in the meeting of 11th November 2002. It shall be recalled that these plots correspond to the residences of the TPA staff and for which the tender for this work was called. The table has also quantified plots corresponding to other properties of TPA and which shall eventually remain under TPA title. The third group of plots are the new ones designed to fill the lands that until now were not developed. It was agreed that the plots be surveyed so that TBA could build more houses for TPA workers also close to the work places.

A: DAR ES SALAAM

Estate	Residential Plots		Other Plots	TOTAL	Comments
	Developed	New			
Msasani 1	30	-	-	30	
Msasani 2	85	2	1	88	
Plot No. 95 Guinea Rd	2	-	-	2	
Plot on Kimweri Rd, Oysterbay	4	-	-	4	
Plot No. 2114/5 Block A Sea View	3	-	-	3	

Plot No. 2104/5 Ocean Rd	2	-	-	2	
Plot No. 269 Alykhan Rd	2	-	-	2	
Plot No. 272 Alykhan Rd	-	-	-	0	
Plot No. 194 Malik/Undali Rd	-	-	-	0	
Plot No. 577 Mindu Street	-	-	-	0	
Plot No. 446 Charambe Street	-	-	-	0	
Plot No. 454 Charambe street	-	-	-	0	
Plot No. 723 Upanga	2	-	-	2	
Plot No. 2445/208 Msimbazi	94	18	1	113	
Mgulani	161	6	1	168	
Plot No219 – 307 Kurasini	130	44	4	178	Excludes Primary School plots
Kigamboni	6	11	-	17	
Plots 1-10 Chang'ombe	10	-	-	10	
Plot No. 43 + 44 Kilwa Rd	4	-	-	4	
Gerezani	-	-	-	0	
Plot 569 Mindu street	8	-	-	8	
	81	7	731		

B: TANGA

Estate	Residential plots	Other Plots	TOTAL

	Developed	New		
House No. 114, Fire Brigade	-	-	-	0
Old usagara Estate Plots 14-26, 10-18, 6-11,1-22 Blocks E, F, G. + H Resp.	90	-	3	93
New usagara Estate Block j Plots 20-56, 58,60,62, 64, 66,67-93,104-107	14	34	1	49
Chuda/Kana Flats 61-63	-	-	-	0
Chuda House No. 65	-	-	-	0
Chuda House No. 29	3	-	-	3
Chuda House No. 30	4	-	-	4
Plot No. 94 Chuda/Chumbageni Block H	22	-	-	22
Plot Nos. 54-56 Chumbageni Block KB VIII	3	-	-	3
Plot Nos. 129,131,133,137,138-147 Chumbageni Block KB VIII	14	-	-	14
Plot No. 27,29,30,31 + 32 Block B Nguvumali	5	-	-	5
Kisosora	6	-	-	6
	34	1	196	

C: MTWARA

S/N	Estate	Residential plots		Other Plots	TOTAL
		Developed	New		
34	Plot 242 Ligula Flat	-	-	-	0
35	Railway Estate	175	92	11	278
Total 175			**92**	**11**	**278**

TP DRAWING EXTRACTS:

S/n	City/ Municipality	Estate	TP Drawing/ Extract No.
1	Dar Es Salaam Kinondoni M. Council.	1. Msasani 1& 2 2. Plot 95 Guinea Rd 3. Plot 56 Kimweri Rd	1/499/169 A 1/503/369F 1/503/369G
2	Dar Es Salaam Ilala M. Council	4. Plot 2114/5 Block A Sea View 5. Plot 2104/5 Ocean Rd 6. Plot 577 Mindu Str 7. Plot 723 Upanga 8. Plot 2445/208 Msimbazi	Consult DHHSD Records
3	Dar Es Salaam Temeke M. Council	9. Mugulani 10. Plot 219-307	1/507/369D at 1:2,500

		Kurasini	
		11. Kigamb oni	
		12. Plot 1- 10 Chang'o mbe	
		13. Plot 43 & 44 Kilwa Rd	
4	Tanga	14. Blocks E, F, G, H & J Usagara	2/TA-12/382/G5A at 1:2,500
		15. Chuda – Kana Block E	2/213/1063/E 5A, at 1:2,500
		16. Plot 94 Block H Chumba geni	2/213/1063/E 5B, at 1:2,500
		17. Plots 27 – 32 Block B Nguvum ali	2/213/1063/E 5C, at 1:2,500
		18. Plots 5, 7, 9, 11 & 13 Block N, Kisosor a	2/213/1063/E 5D, at 1:2,500
5.	Mtwara	19. Railway	3/97/369

REVISION OF CONTRACT SUM:

<u>A: New Work Load and Cost Thereof:</u>
The tripartite meeting of 11th November 2005 agreed
on a number of issues that constitute additional
workload for the consultant. These decisions are:
1. To include non-residential properties in the
   current exercise.
2. To engage a town – planning professional for
   the design of layouts covering the undeveloped
   areas on the estates.
3. To process through TBA and HSDD the
   approval of the TP-Drawings designed and
   drawn by the town planner.
4. To include the new layouts in the survey so that
   TBA may build houses for other workers of the
   TPA.

The layout plans are being forwarded to DHSD
through TBA for approval. There are 226 new plots
(Both residential and others). The inception report
revealed that another 127 plots in Tanga had not been
listed in the tender documents. The total new number
of plots to be surveyed are therefore (226+127) three
hundred fifty-three (353) plots.

The additional workload above is costed in the table
below:

S/N	Item	Unit Cost	Amount
1	226 Newly created plots	70,000/=	15,820,000/=
2	127 plots omitted in tender documents	70,000/=	8,890,000/=
3	Sub – Total 1		24,710,000/=
4	Cost of designing layouts (12% of sub – total 1)		2,965,200/=
5	Sub – Total 2		27,675,200/=
6	VAT (20% of Sub-Total 2)		5,535,040/=
7	**TOTAL**		**33,210,240**

Rates used in calculating above cost are same as used in Original Tender.

B: Already Agreed and Contracted Work Load:
Contract Sum as per agreement of the 31st August 2005 is sustained

C: Variation to Contract Sum
Amount equal to cost of new work load i.e., TShs 33,210,240.00 or thirty-three million, two hundred and ten thousand, two hundred and forty shillings only.

The Director

     <u>27th October, 2005</u>

Engineering and Technical Services Dept.
Tanzania Ports Authority
P. O. Box 9184
Dar Es Salaam

Dear Sir,

Re: Cadastral Survey of Former TPA Housing Estates at DSM, Tanga and
    Mtwara.

Subject: Implication of Surveying Non – Residential Plots and Dividing Open Spaces into Plots.

Please refer to the subject above. The Inception Report submitted to you on 12th October 2005 brings to your attention the consequences of dividing land under title. In particular, we have impressed upon you the fate of Tenure security existing non-residential properties and open spaces on these estates.

At a meeting on Monday 24th between Clients coordinator and Dr. Lugoe, The Consultant was required among others, to submit to you the cost implications of the consequences of cancelling the existing survey plans and titles. Such cancellation, we add, is a necessary step in allowing a re-survey and individualization of title.

Herewith is a schedule of additional work evolving as a result of the cancellations. It is additional work because the project was focused on residential houses sold to TPA workers and not more. We have also provided the cost estimates based on rates of 70,000/= per plot as used in our submitted tender (page10) and accepted by your Tender Board. We have included a 12% mark-up for planning the open spaces, which is a requirement of the Town Planning Offices. We hope this enables you to make headway on section (5), item (2) of the inception report.

As a further qualification on section (3.3), bullet (3) of the same report please be in formed that Government now disposes each plot at an average rate of 1,000/= per sq m. and prime land such as your Kigamboni would be at a high premium enough to pay for the cost of this project and create a huge surplus.

I remain

Yours Sincerely

Dr. F.N. Lugoe

<u>Managing Director</u>

SUBMISSIONS

Director of Engineering and Technical Services
Tanzania Harbours Authority
P. O. Box 9184
Dar Es Salaam, Tanzania.
                    January 17th 2007

Attention: CERS, Mr. Safari

Dear Sir,
**Re: Cadastral Surveying of Former TPA
Housing Estates at
        Dar Es Salaam. Tanga and Mtwara.**

Subject: Progress

Further to our telephone conversation of Monday 15th
January, 2007, Please note the following:
- o The municipalities have now, to various
    degrees, responded to the request of TBA made
    in January 2006, to approve TP Drawing
    extracts of the estates under this project. Topo-
    Carto Consultants Ltd submitted these to the
    Director of Housing and Human Settlements
    Development (DHHSD) for approval on 18th
    October 2006 vide a covering letter copied to
    you.

- o Drawings for Tanga City and the municipalities of Kinondoni and Mtwara have subsequently been approved by the DHHSD and are now being used by the Director of Surveys and Mapping to finalise our work that was submitted as per our letter of 25th July 2006 to you.
- o Drawings for Ilala have been submitted to DHHSD but are yet to receive due blessings of the supreme planning authority.
- o Drawings for Temeke Municipality have been split into two parts, namely; Mugulani and Others. The TP Drawing for Mugulani Estate has been approved by the DHHSD, whilst the rest have not been submitted to the UPC at Municipal Hall for consideration. The Municipal Town Planner was transferred to Musoma in November 2006. His place has only recently been filled but handing over/ taking-over is only expected to take place towards the end of this month.
- o Please find, herewith, copies of the approved TP Drawings for your perusal and records.

Yours Truly,

Dr. F. Lugoe
CEO & Lead Consultant.

First Submission:

Director of Engineering and Technical Services
Tanzania Harbours Authority
P. O. Box 9184
Dar Es Salaam
Tanzania.

April 2, 2007

Attention: CERS, Mr. Safari

Dear Sir,

Re: Cadastral Surveying of Former TPA Housing Estates at
Dar Es Salaam, Tanga and Mtwara.

Subject: Submission of Registered Survey Plans for Msasani and Mtwara Estates

I am forwarding to you copies of survey plans in the above captioned project that are duly approved and registered under the Land Survey Ordinance (CAP 390) by the Director of Surveys and Mapping. The particulars are entered in the table below:

S/	Estate	Plan	Registe	PLOTS

n		Number	red Plan Number	Total Number	Serial Numbers
1	Msasani 1, Kinondoni Municipality on TP Drawing No. 1/499/169 A	E¹ 210/107	45304	30	From: 1829/1 To: 1829/30
2	Msasani 2, Kinondoni Municipality on TP Drawing No. 1/499/169 A	E¹ 210/107	45304	89	From: 1826/1 To: 1826/89
3	Railway Estate, Mtwara	D²¹ 28/4	45221	191	From: 27 To:

					217
	Municipality on TP Drawing No. 3/97/369				
4	TOTAL			310	

Yours Truly,

signed
Dr. F. Lugoe
CEO & Lead Consultant.

Copy to: Tanzania Buildings Agency

Second Submission:

Director
Engineering and Technical Services
Tanzania Harbours Authority
P. O. Box 9184
Dar Es Salaam.
Tanzania.

September 10 2007
Attention: CERS, Mr. Safari

Dear Sir,

**Re: Cadastral Surveying of Former TPA Housing Estates at
     Dar Es Salaam, Tanga and Mtwara.**

Subject: Submission of Registered Survey Plans for Various Estates

I am forwarding to you copies of survey plans for cadastral surveys undertaken in various estates of the above captioned project that are duly approved and registered under the Land Survey Ordinance (CAP 390) by the Director of Surveys and Mapping. Particulars of each Approved Survey, per Block, are entered in the table attached to this letter.

Surveying of Former TPA Housing Estates at Dar Es Salaam, Tanga and Mtwara.

The first submission of 310 for the two estates in Msasani, Dar Es Salaam and the Railway Estate in Mtwara was made to you vide my latter of 2nd April 2007.

I remain,

Yours Truly,

Dr. F. Lugoe
CEO & Lead Consultant.

Copy to: Tanzania Buildings Agency
With attachments

Second Submission: Completed, Approved and Registered Cadastral Survey Plans for Various Estates in Tanga City and Ilala Municipality

Third Submission:

The Director
Engineering and Technical Services
Tanzania Harbours Authority
P. O. Box 9184
Dar Es Salaam.
Tanzania.
4th April 2008
Attention: CERS, Mr. Safari

Dear Sir,

Re: Cadastral Surveying of Former TPA Housing Estates at
**   Dar Es Salaam, Tanga and Mtwara.**

Subject: Submission of Registered Survey Plans for Various Estates

Please receive copies of survey plans for cadastral surveys undertaken in various estates of the above captioned project that are duly approved and registered under the Land Survey Ordinance (CAP 390) by the Director of Surveys and Mapping. Particulars of each Approved Survey, per Block, are entered in the table attached to this letter.

The first submission of 310 for the two estates in Msasani, Dar Es Salaam and the Railway Estate in Mtwara was made to you vide our letter of 2nd April 2007.

The second submission of 123 plots in nine estates of Tanga, and Ilala Municipality was made to you through our letter of 10th September, 2007

This submission of 184 plots in Tanga, Kinondoni and Temeke Municipalities brings therefore the total number of plots approved by the Director of Surveys and Mapping to 617. All 617 plots were ready for titling on the dates of approval and registration of the respective survey plan.

I remain,
Yours Truly,

Dr. F. Lugoe
CEO & Lead Consultant.
Copy to: Tanzania Buildings Agency
Enclosure: Copies of Registered Survey Plans attached

Fourth Submission:

Estate	Plan Number	Registered Plan Number	PLOTS	
			Total	Numbers
Chumbageni Block "H"	D² 317/7	50550	18	94/1 – 94/18
Chang'ombe Block "C"	D' 510/54	50847	10	1-10
Oysterbay	E' 210/112	50549	2	95/1 – 95/2
Kurasini	D' 506/53	50484	139	219/1 - 307
Kurasini	D' 500/4	47953	4	43/1 – 44/2
Kigamboni Block "D"	E' 221/8	50845	11	1 - 11
TOTAL			184	

The Director

Engineering and Technical Services
Tanzania Ports Authority
P. O. Box 9184
Dar Es Salaam, Tanzania.

22nd April 2008

Attention: CERS, Mr. Safari

Dear Sir,

**Re: Cadastral Surveying of Former TPA
Housing Estates at
       Dar Es Salaam, Tanga and Mtwara.**

**Subject: Submission of Registered Survey Plans
for Mgulani Estates**

Please receive copy of survey plans for Plots Nos. 602
– 775 of Mgulani Estate in Temeke Municipality, in
partial fulfillment of our duty in the above captioned
project. The approved survey has 174 plots on Plan
No. D' 506/52 and Registered Plan No. 51100.

This is our fourth submission that brings the total
number of plots surveyed and approved to 974
distributed as follows: (i) The first submission of 310
for the two estates in Msasani, Dar Es Salaam and the
Railway Estate in Mtwara was made to you vide our
letter of 2nd April 2007; (ii) The second submission of

123 plots in nine estates of Tanga, and Ilala Municipality was made to you through our letter of 10th September, 2007; (iii) The third submission of 183 plots in six estates of Tanga City, Kinondoni and Temeke Municipalities was made through our letter of 4th April, 2008; and (iv) today's submission of 184 plots in Tanga, Kinondoni and Temeke.

All plots are ready for titling on the basis of the respective registered survey plans. This submission completes our work in Temeke Municipality as we have done in Kinondoni, and Mtwara Municipalities. Still awaited from the Director of Surveys and Mapping are approvals of surveys of 16 plots for Kisosora (6) in Tanga, and Plot 577 (8) and 723 (2) in Upanga, Ilala Municipality.

Please allow us to submit our third invoice for payment under this project of 20% of the agreed sum that takes into account agreed variations to the contract sum. Also please note that plots for vacant areas in estates of Kinondoni, Temeka and Ilala Municipalities have been approved with the approval of the survey plans but these, and expenses incurred in payments for allowances of Municipal Councillors, will be claimed upon your final certification.
I remain,
Yours Truly,

Dr. F. Lugoe
CEO & Lead Consultant.
Copy to: Tanzania Buildings Agency
Enclosure: Copies of Registered Survey Plan attached

SUBMISSION SUMMARY

PROJECT: THE SURVEY OF TPA STAFF HOUSES

Objective:
The tender documents define the task ahead as one of cadastral surveying of former TPA estates to enable each property owner to be issued with a letter of offer of a right of occupancy and later on, a certificate of occupancy.

After the houses in these estates were sold off to TPA staff in October of 2003, it became apparent that the clustering of houses did not render itself to individualization of title. It can be seen that many houses are clustered in residential housing estates whilst many others are flats or condominiums located on land with a common title in the name of the Tanzania Ports Authority and most importantly is the fact that the majority of houses are semi-detached and for which the individualization of title will only be possible with change in attitude of authorities regarding land parcel definition for semi-detached houses..

General Approach to Cadastral Surveys:

A visit to all sites revealed the following scenarios:

1. Estates that are a subject of this exercise contain more than residential properties. In other words there are properties on these estates that have not been disposed away and remain the properties of TPA or Tanzania Buildings Agency, TBA. Among these landed properties are; nursery schools, social clubs, sewers and sewerage plants (existing or planned), signal towers, play grounds, etc. The consultant needs clear guidance as to what should be done under the circumstance. The practice is that no titles can be granted within an existing title. Conversely, the Director of the Surveys and Mapping Division will not approve a cadastral survey within an approved survey. An Alternative is to aim at a cancellation of the registered survey plan of the estate and hence a revocation of its title and resurvey all the properties including open spaces for subsequent titling to the new owners.

2. Following from (1) above also is that TPA seems to have no other options save to survey all properties and the open spaces in its estates. The practice is that each land parcel (urban plot) must have access to a public right of way. All streets within an estate are private to that estate. These will only become public either upon cancellation of the registered survey plan or upon a resurvey that re-defines the road/street network. Although cancellation of the survey plans may seem easy and straight way out. TPA should weigh other consequential outcomes before proceeding with this alternative. Advice will be given here after discussing the properties briefly here under.

3. Three structural categories of properties on the estates have been identified.

   - Some of the properties such as single residential houses and maisonettes (ground floor and those above it comprise a single unit), are easy to survey for individualized title, for the simple reason that each unit occupies a unique part of the land.

   - A second category in residential flats comprises of vertically adjoined units of two or more flats with a common entrance for the ground and higher floors. In this case individualised titles cannot be framed for each housing flat but for the unit of several flats. This shortfall is brought about because Tanzania does not have a legislation providing for strata titles. In such a case only co-occupation is possible and the land and property shall be occupied in common (see Part XII of the Land Act No. 4 of 1999) by all who have purchased property in a given unit of a vertically stratified block of flats. It is proposed that the subdivision of land be done along the lines separating the different units so as to reduce the number of co-occupiers in a common title to a minimum.

- A third category in residential flats is all those flatted buildings, which cannot be apportioned into logical units of fewer individual co-owners. Most of these have a common entrance and a common corridor on each floor above the ground floor. Again, in the absence of strata titles, the provisions for co-ownership in the Land laws are applicable and will apply to all who have purchased flats in each building block.

4. The Client is well advised against relinquishing its land claims by cancellation of the registered

survey plan. Besides the properties spelled out under (1), several other consequences ought to be seriously considered including;

- Relinquishing claims to land ownership prior to extinguishing all currently held interests means that any vacant land on the estate reverts to the planning authority, which in this case is either the Municipality or the City Council. Experience has shown that immediately after it becomes public knowledge that TPA has pulled out, there will be land-grabbing worse than what has happened to the Msimbazi Estate, so far.

- Further, since the vacant land to be left behind as a result of relinquishing land interests prematurely is unplanned, then any subsequent development by land grabbers will deteriorate into squatter land and the Client will be blamed for not being visionary enough as to prevent such a situation from occurring.

- Finally, and as a consequence to the TPA workers their properties if surrounded by shanties will lose value and could lead to court cases with compensation claims.

5. Proper land disposition is one that finds a legal owner as and when the Client's land interests in the estates are extinguished. In this case all land

on the estates including open spaces and non-residential properties must find legal owners. This is possible if all the land is surveyed at the time of subdividing the estates for transfer of ownership of residential properties to TPA workers. Such an approach will redefine the public rights of way that are not encumbered by TPAs original land title and also vest the rest of the non-residential properties to self, pending another phase of disposition should that be in the interest of the Client. The large open spaces may be surveyed as commercial and additional residential plots to be allocated to staff who may be landless but were not offered any land out of the Client's stock.

PROJECT EXAMPLE NO.5: PROJECT TO GENERATE A TOPOGRAPHICAL MODEL OF DANGOTE SITES IN MTWARA BY LAND SURVEYING METHODOLOGY

ENGAGEMENT:

DANGOTE INDUSTRIES LIMITED, TANZANIA

Bray Properties, House No.5, Plot 80,
Next to Swiss Embassy, Kinondoni,
P.O.Box 12038, Dar Es Salaam,
TANZANIA

DIL/VSD/PROJ/TZ/002/2011
Dated: 02.03.2011

To,

Topo Carto Consultants Ltd,
P.O. Box 35619,
Dar es Salaam, Tanzania.

Sub.: Scope of work to carry out a Topographical Survey of proposed Cement Plant site and Terminals for Dangote Cement Works Ltd.

Dear Dr.Furaha,

CEO & Lead Consultant,

With reference to our communication on e mail and your financials already submitted for the job, we would like to inform you the scope of work for the **Topographical Survey of proposed Cement Plant site at Mbuo, Export Terminal in Mtwara Port & Import Terminal in Dar Port.** You are requested to kindly go through it and revise your offer and send again to us.

A.   Scope of Work:

1.   Plant area topographical mapping of (1) Main Plant : the approximate area 150 hectors (2) Export Terminal in Mtwara Port : Area 30,000 SQM (3) Import Terminal in Dar Port : Area 6000 SQM.

2.   The mapping is required at a scale 1:1000 with contour intervals of 0.50 to 1.0m intervals.

3.   Perimeter survey should be carried out with the view of confirming that the area of land and perimeter shown on the survey plan drawing agrees with the site boundary.

4.   All the topographical features like hills, valley's, escarpments, spurs, saddles,

plateaus, forest, planes, sea, lakes, rivers, ponds, cannels, swamps, towns, railways, roads, embankments, boundaries and other prominent features like church, mosque if any etc.,

5. All documents should be provided in electronic format along with hard copies.

B. Contractors Other Responsibilities :

1. This will start with a general reconnaissance to locate existing Government Registered Survey Controls points and also to establish the actual extent of the project area.

2. The Survey will be connected to the National Grid.

3. Because of the sensitive nature of the Project, all necessary procedures and precautions that ensure the utmost accuracy will be used.

4. Boundary lines will be cleared mechanically, and beacons that will match the nature of terrain shall be established at every change of direction along the boundary of the site.

5. Horizontal and Vertical Co-ordinates i.e.(x,y.z) of survey points would be accurately determined to serve as Permanent Bench marks

during Survey work for design and construction reference.

6. Relevant information and permits required will be officially collected from the survey department of Ogun State and relevant authorities by the contractor.

7. Connecting of the survey to the existing controls.

8. Establishing of permanent survey beacons at every 25 meters along the centre

9. Provide spot heights (levels) at every 10 meters interval and cross sections of

   10 meters either side of the centre line.

10. Produce topographical survey with either 0.50 to 1.0m contour of the route.

11. Produce a digital drawing (plan/profile) of the area.

C.   Deliverable:

i) Drawing will be produced as specified scale
ii) Four (4) copy of the digital plan drawing at specified scale.
iii) Survey technical reports
iv) One CD containing item (ii&iii) above

D.   Work Schedule:

This job is to be completed in Four to Six weeks.

As I informed you, the rate quoted by you is very much on higher side, You are requested to send your revised commercial proposals.

We look forward for a mutually beneficial relationship.

Yours faithfully,
For **Dangote Industries Limited, Tanzania**

VIDYA SAGAR DIXIT
(Senior General Manager - Projects)

Mr. V. S. Dixit
Sr. General Manager - Projects,
Dangote Cement Works Ltd
Plot No. 80, House No. 5, Kinondoni Road,
P. O. Box 12038, Dar Es Salaam,
Tanzania.
8th March 2011

Your Reference: DIL/VSD/PROJ/TZ/002/2011
Dear Sir,

Re: Revised Quotation for Topographical Survey Works of Proposed Cement Plant Site at Mbou in Mtwara and Cement Terminals in Mtwara and Dar Es Salaam Ports

After reviewing our quotation of 6th March 2011 with you this afternoon and in response to your further request for a revised quotation we hereby note the following agreed changes and revise the quotation with full consideration of these changes:

1. The area for the Cement plant has now been confirmed to be 1,500 hectares and not 150 hectares. Topo-Carto Consultants Ltd has proposed to possible client to limit the topographical survey to an area of about 150 ha as the actual location of the plant. The remaining will be secured by a well marked

boundary survey and probably developed as an environmental management show case by planting trees or similar developments for the remaining area so as to limit intrusion and land conflicts with neighbours. The boundary survey would therefore be more involving and carries with it cost implications. The area under survey would remain as earlier proposed and would therefore not have cost implications.

2. Client expects to extend the area of development at the the Mtwara port terminal to the ocean, which could increase the area by about 25-30 percent. The request to include the additional area has cost implications and is therefore considered in this revision.

3. Topo-Carto Consultants has a stock of digital aerial photo mosaics of Dar Es Salaam in full colour and will endeavour to superimpose the topographical survey on to the mosaic if the plotting scale and resolution will allow. This attempt has no cost implications in this quotation.

4. In this revision Topo-carto Consultants will also consider unit costs for one hectare of topographical surveys at two different scales namely: 0.5 – 1.0 m and 1.0 – 2.5 m

5. In anticipation of the project being awarded to us, we make further downward adjustments on the proposed contract sum for both works in

Mtwara by removing the double mobilizations and demobilizations. In this regard the completion time has also been re-considered.

QUOTATION:

We now offer to execute the works on the three sites in Dar Es Salaam port, Mtwara port and the main plant site in Mbou village, Mtwara District on the following terms.

- The works on the Mtwara sites will be carried out within two months commencing work at the same time on both works. Topo-Carto Consultants will start with technical works at the Mtwara Terminal at the same time that line clearing works at the Main plant also commences. The work at the Dar Es Salaam port terminal will take two weeks. In all cases rainy days are not to be counted but will constitute a *force majeur*.

- Producing one topographical plan at 1:1,000 scale with 0.5-1.0m contour interval for each site except where the client commissions extra works at different and specified scales.

- The amounts quoted in this submission are exclusive of the Value Added Tax (VAT) at 18%.

Site	Area size (in	This quotation	Comment

	hectares)	(in USD)	
Cement Plant at Mbuo	1,500.0	35,250.00	Full perimeter surveyed
Mtwara Export Terminal	4.0	7,000.00	Includes two-way mobilizations of $1,200 that could be deleted if client grants both sites in Mtwara
Dar Import Terminal	0.6	4,500.00	Aerial photo mosaics to be used as stated earlier

The quotation for the 1,500 ha Main Cement Plant topographical works at Mbuo is Thirty Five thousand Two Hundred and Fifty ($ 35,250.00) USA dollars only.

The quotation for the Mtwara Export Terminal topographical works is Seven Thousand USA dollars and forty cents ($ 7,000.00) only.

The quotation for the Dar Es Salaam Import Terminal topographical works is Four Thousand Nine Hundred and Fifty Six ($ 4,500.00) USA dollars only.

ADDITIONAL WORKS:

Additional works commissioned by the client will be at the following rates:

1. One hundred and fifty ($ 150.00) USA Dollars per hectare of topographical survey plotted at 0.5-1.0 m contour interval.
2. Ninety ($ 90.00) USA Dollars per hectare of topographical survey plotted at 1.0-2.5 m contour interval.

TECHNOLOGY:

All works in the topographical surveys quoted will be done using three basic technologies: (i) the Global Positioning System (GPS) operating a set of three ProMac 3 satellite receivers and data processing system; (ii) the Total Station system for digital 3D (X, Y, Z) position fixation; and (iii) Spirit levelling for precise determination of elevations. Please see our list of equipment and technology in the table below.

Item of equipment	Description, Make, Year and Age (years)	Condition and number available
GPS Receiver sets	Ashtech Pro-MAC 3, 2008 Omnistar GPS	Set of 3 in very good condition
Total Station	Topcon GTS 700 complete with accessories & 2 sets of Prisms, 2005	1 set in good condition
Total Station	Sokkia SET 4 complete with accessories, 2002	1 set in good condition
Electronic	Sokkia DT 5 With	1 set in good

Theodolite	accessories, 2005	condition
GPS Receiver Unit	Garmin GPS 72 HHGPS Satellite Receiver, 2005	1 unit in good condition
Desk-Top Computer	Mercury Pentium4 (2004) with DVD/CD-RW, 2.5GHz, 120 GB HDD.	3 sets in good condition
Desk-Top Computer	HP Pentium3 (upgrade) with 512 MHz 80 GB HDD (2004)	1 set in good condition
Lap-Top Computer	Acer TravelMate 252LC_DT DELL Latitude D610	2 units in good condition
Computer Printers	1. HP 2002 A3/A4 Colour Printer (2006) 2. HP 3600 Series A4 Colour Printer (2007) 3. HP 1010 A4 B/W Laser Printer (2007) 4. HP 1018 A4 B/W Laser Printer (2008) 5. Cannon IR 1800 A4 B/W	1 unit each in good condition
OS Computer Software (basic)	Windows XP, MS-Office XP & 2003 incl. MS project, spreadsheet, dbase, Adobe Acrobat & MS-Word., etc.	New
Professional Computer Software	1.AutoCad R14, 2002 & 2008 2.Geomedia v.4, (20050 3.Arc GIS (2005) 4. Planet GIS, (2005) 5.CadPro (2004 installation)	New
Vehicles	1. Pajero Turbo-Wagon 4WD 2. Toyota MarkII Sedan 3. VW Transporter	Very Good
Communication Devices	1.CB Radios (10) short & long range 2.Cellular phones 3. fax machine (2773631)	Very good
Secretarial Equipment	1. Mustek flatbed Scanner; 2. Cannon IR1080 printer scanner and photocopy machine (2008); 3. Binding Machine	Very good

| | 4. guillotine and
5. lamination machine, etc. | |

This is a revised submission. All other information submitted to you including; the Scope of Work, Company Profile, References to Projects Undertaken, Curriculum Vitae and List of key staff remain as per yesterday's submission.

SIGNATURE

Dr. Furaha N. Lugoe, flugoe@yahoo.co.uk
CEO and Lead Consultant
Topo-Carto Consultants Limited

Appendix A – Schedule of Payments and reporting Requirements

Milestone	Product	Payments
Signing of Contract and Mobilisation (Advance payment)	Contract in Place	50%
Submission of topographical map and data for the project	Topographical Map	50%

Appendix B – Key Personnel For Project Execution

Position	Name	Years of experience
Management: 1) Chief Executive	F. N. Lugoe	33
Administration: 1)Business Management	G. Lugoe	20
Surveyors: 1)Surveyor 2) Field Surveyor	F.N. Lugoe A. A. Msuya	33 22
1) GIS & Mapping	A. Kingu	8
Other: 1)Field Assist/Driver 2)Field Assist/Driver 3)Field Assist. 5)Field Assist.	C. S. Ndaki E. Kipemba G. Waito A. Mchomvu	7 10 6 4

Revised Quotation

Site	Area size (in hectares)	This quotation (in USD)	Comment
Cement Plant at Mbuo	150	27,842.00	
Mtwara Export Terminal	40,000 m²	5,841.60	Mobilizations excluded
Dar Import Terminal	6,000 m²	5,167.20	

Breakdown of Quotation

s/n	Site	Area size	Activity	Quantity	Unit rate ($)	Amount ($)
1	Cement Plant at Mbuo	150 ha.	Mobilisation	2 trips	500	1,000
			Reconnaissance	1 day	120	120
			Bush clearing	15 days	30	450
			Beaconing	4	25	100

			of survey control & elevation points	pillars 15 IPC 2500 pegs	2 0. 5	30 1250
			Establishi ng Survey Control both horizontal and vertical	19 in 2D 15 in 1D	50 15	950 225
			Connectin g Survey control to national network	1 traver se 1 L/Lin e	65 0 15 0	650 150
			Boundary survey	2 traver ses	65 0	1,300
			Survey of grid points (spot heights)	2400 points in batch es of 4	8	4,800
			Survey of detail	30 points	5	150

s/n	site	Area size	Activity	Quantity	Unit rate ($)	Amount ($)
			Cartographic work (contouring & plotting of detail)	lumpsum		9,000
			Validation (Field Checking)	3 days	120	360
			Administrative cost	lumpsum		2,500
	SUBTOTAL					24,035.00
			20% profit			4,807.00
	TOTAL (1)					**27,842.00**
2	Mtwara export terminal	40,000 m² OR 3 ha.	Mobilisation*			
			Reconnaissance	1 day	120	120
			Beaconing	4 pillars	252	10010

					5 IPC 16 pegs	0.5	8
				Establishing survey control both horizontal and vertical	9 pts in 2D 6 in 1D	50	450 90
				Boundary survey	1 traverse	650	650
				Survey of grid points (spot heights)	64 points	5	320
				Survey of detail (detail picking)	20 points	10	200
				Cartographic work (contouring and plotting of detail)	lumpsum		2,200
				Validation (Field	1 day	120	120

s/n	site	Area size	Activity	Quantity	Unit rate ($)	Amount ($)
			Checking) **			
			Administrative cost	lumps um		600
	SUBTOTAL					4,868.00
			20% profit			973.60
	TOTAL (2)					**5,841.60**
s/n	site	Area size	Activity	Quantity	Unit rate ($)	Amount ($)
3	Dar Import Terminal	6,000 m² OR 0.6 ha.	Mobilisation	-		-
			Reconnaissance	1 day	120	120
			Beaconing	4 pillars 5 IPC 10 pegs	25 2 0.5	100 10 5
			Establishing Survey Control	10 pts in 2D 10 in 1D	50 15	500 150

			horizontal and vertical			
			Boundary survey	1 traverse	650	650
			Survey of grid points (spot heights)	10 points	5	50
			Survey of detail (detail picking)	20 points	10	200
			Cartographic work (contouring and plotting of detail)	lumpsum		2,000
			Validation (Field checking) **	1 day	120	120
			Administrative cost	lumpsum		400
	SUBTOTAL					4,306.00
			20% profit			861.2

					0
TOTAL (3)					**5,167. 20**

* excluded on the assumption that both Mtwara Sites are commissioned to TCC.

**includes travel back to site with client – client to provide two way transport

TOPOGRAPHICAL MAP OF DANGOTE MAIN CEMENT PLANT SITE IN MTWARA DISTRICT, TANZANIA
LEGEND
Contours
Water tank
Control points
Sparse Forest
Thick Forest
Iron Pin in Concrete (IPC)
Spot Height Point
SHEET HISTORY
COMPILED IN JAN' 2012
SURVEYED AND DRAWN BY:
TOPO-CARTO CONSULTANTS LTD
topocarto@raha.com
MEASUREMENTS ARE IN METRES
CONTOUR INTERVAL : 1 METRE
GRID DATA
Grid: U.T.M. Zone 37
Datum: Arc 1960
LILUNGU 296/X/7
KDQ 799
Area = 156.055 hectares
SCALE 1:2000
LINEAR SCALE (IN METRES)

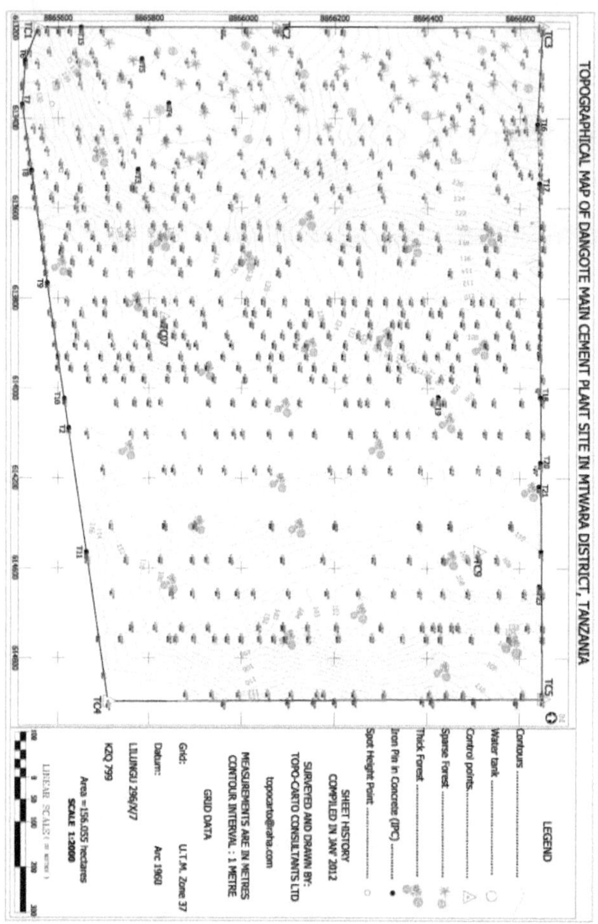

END

About the Author:

Dr. Furaha Lugoe is author of over 120 publications in books, journal papers, policy briefs and conference proceedings. In academia, Dr. Lugoe gained a 15 year-experience in teaching and research in the fields of geomatics engineering and geodesy at the Universities of Dar-Es-salaam (UDSM), New Brunswick (UNB), Zimbabwe (UZ) and the now Ardhi University, with links to universities in UK (UCL, North-East London, New Castle upon Tyne), Scotland (Glasgow) and Germany (Hanover). Dr. Furaha Lugoe has undertaken many land surveying projects in his lifetime. These projects cover areas of land divestiture, dredging, route location, land titling, structure deformations, land-take, and terrain modelling - from Chernobyl (1977) to North Mara Gold Mines (2014). Further, for over 20 years now, Dr. Lugoe has been an active and reliable consultant and advisor in land administration, land governance, ecosystem management and human settlements, focused on policy making and assessment and design of strategies (2000 – 2017).

www.ingramcontent.com/pod-product-compliance
Lightning Source LLC
Chambersburg PA
CBHW020310160726
47992CB00004B/1472